T0257779

# Sourcebook of Numerical Modeling of Mass Transfer

Edited by **Preston Runner**

CLANRYE INTERNATIONAL

New Jersey

Published by Clanrye International,
55 Van Reypen Street,
Jersey City, NJ 07306, USA
www.clanryeinternational.com

Sourcebook of Numerical Modeling of Mass Transfer
Edited by Preston Runner

International Standard Book Number: 978-1-63240-470-1 (Hardback)

# Contents

**Permissions**

**List of Contributors**

# Preface

This book has been an outcome of determined endeavour from a group of educationists in the field. The primary objective was to involve a broad spectrum of professionals from diverse cultural background involved in the field for developing new researches. The book not only targets students but also scholars pursuing higher research for further enhancement of the theoretical and practical applications of the subject.

This book presents beneficial research into advanced numerical modelings. The expertise of mass transfer processes has been extended and applied to different realm of science and engineering including industrial applications in recent years. Since mass transfer is primeval phenomenon, it plays a vital role in the scientific researches and fields of mechanical, energy, environmental, materials, bio and chemical engineering. In this book, experts provide recent developments in scientific findings and technologies, and introduce new theoretical models concerning mass transfer for sustainable energy and environment. Thus, the book provides beneficial references for research engineers working in the variety of mass transfer sciences and related fields, and also be resourceful for university students.

It was an honour to edit such a profound book and also a challenging task to compile and examine all the relevant data for accuracy and originality. I wish to acknowledge the efforts of the contributors for submitting such brilliant and diverse chapters in the field and for endlessly working for the completion of the book. Last, but not the least; I thank my family for being a constant source of support in all my research endeavours.

**Editor**

# Advanced Numerical Modelings

# The Theory of Random Transformation of Dispersed Matter

Marek Solecki

Additional information is available at the end of the chapter

## 1. Introduction

The development of civilization, with regard to its consequences, requires appropriate research tools. This is the reason of significant progress in different fields in mathematical modelling of processes involving mass transfer. For their description there are often used widely known models which can be written in a general form of nonlinear differential equations. Woltera [1], who conducted the research of oscillation level of selected fish species in the Adriatic, described the interaction system in the population of predator-victim type. It is compatible with the obtained by Lotka [2] description of the reaction with the expected oscillations in the concentrations of chemical compounds. The process of destruction of organisms during disinfection is included in the model of Chick [3] developed by Watson [4]. The kinetics of the course of simple enzymatic reactions was described by Michaelis and Menten [5]. Mathematical description of the process of microorganisms disintegration in high-pressure homogenizer was developed on the basis of experimental data by Hetherington et al. [6]. After over twenty years since revealing the effect of the position of enzymes in a cell on their release rate [7], Melendres et al. [8] proposed a nonlinear description of the process as a consequence of the following events: cell disruption and release of intracellular compounds. The results of pioneering protease inhibitor therapy in HIV infection were the basis for the development by Ho et al. [9] of a simple model of treatment and solving the mystery associated with the quasi-stationary phase of infection. These models are still used to describe the processes or are often the basis for future studies of more complex mathematical problems.

The essence of all the above processes is random transformation of material objects dispersed in a limited space. Their relationships may help to increase the pace of knowledge development. A uniform theory of the presented issues was developed by Solecki [10]. The common general concept based on the knowledge and understanding of important factors shaping the specified sphere of events will facilitate particularistic analysis as well as transfer of knowledge and experience. The aim of the presented study was to develop a unified theory of the presented issues, allowing for phenomenal and mathematical modeling of various processes based on mass transfer.

## 2. Paradigm

This section presents a coherent conceptual system of uniform theory which includes various processes of random transformation of dispersed matter in a limited space.

### 2.1. Space and material objects

Let there be a set $N$ consisting of $n$ elements that are material objects. We assume that $n$ is a natural number.

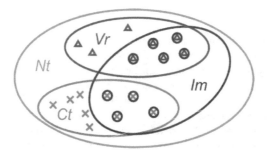

**Figure 1.** The set of identified properties of a material object belonging to set $N$.

Identified properties of material objects belonging to set $N$ are the elements of set $Nt$ (*notum*) (Fig. 1). The property of an object is the feature which characterizes it. Objects can have $\chi$ same properties ($ct_1$, ..., $ct_\chi$) belonging to set $Ct$ (*constans*) which is a subset of set $Nt$. Objects from set $N$ can differ with $\xi$ properties ($vr_1$, ..., $vr_\xi$). They belong to set $Vr$ (*variabilis*) which is a subset of identified features of object $Nt$. We assume that the $\psi$- th property $vr_{\psi i}$ of any $i$-th object belonging to set $N$ is included within the range described by the relationship

$$vr_{\psi\,\min} \leq vr_{\psi i} \leq vr_{\psi\,\max}. \tag{1}$$

e feature of an object belonging to set $Vr$ involves a change of the parameter $vr_{\psi i}$ y the relationship (1). An object from set $N$ may lose some features from set $Vr$ or ment belonging to set $N$ must have all the $\varphi$ features belonging to set $Im$ (*inamissi-* y are the basic features of any object from set $N$ and satisfy the relationship

$$0 < \varphi \leq \chi + \xi. \tag{2}$$

Set $Im$ is a subset of set $Nt$. We assume that after the loss of at least one basic feature of set $Im$ the object is no longer what it was, that is it no longer belongs to set $N$.

Objects from set $N$ have limited duration after which they lose at least one basic feature from set $Im$ and they do not belong to set $N$ anymore.

It is assumed that the fact of existing of objects in set $N$ results in a possibility of generating new elements which are objects of set $N$.

Let there be a limited medium of volume $V$ described by the relationship according to the equation

$$V = f_1(t). \tag{3}$$

In his medium there is set $N$ consisting of $n$ dispersed material objects described by the relationship

$$n = f_2(t). \tag{4}$$

We assume that the volume of material object $V$ is incomparably greater than volume $V_{ni}$ of any $i$-th object of set $N$, according to the formula

$$V_{ni} \ll V. \tag{5}$$

At the initial moment $t_0 = 0$ the number of objects from set $N$ dispersed in space $V$ amounts $n_0$, where $n_0$ is large natural number. We assume that after time $\Delta t$ the number of objects belonging to set $N$ may be changed. Changes may be caused by:

• addition of new objects from the outside to volume $V$ – introduction,

• exclusion of some objects from volume $V$ to the outside – removal,

• generating new objects resulting from the fact of existing of material objects from set $N$ in volume $V$ – multiplication,

• natural exclusion of objects from set $N$, resulting from its duration appropriate for the elements of this set, caused by the loss of at least one feature from set $Im$ – exclusion.

Change in number of objects belonging to set $N$, resulting from introduction, removal, multiplication and exclusion, described by functions respectively $f_{31}(t)$, $f_{32}(n, t)$, $f_{33}(n, t)$ an $f_{34}(n, t)$, at the moment $t$ is described by the relationship consistent with the formula

$$\Delta n_A = f_{31}(t) - f_{32}(n,t) + f_{33}(n,t) - f_{34}(n,t).\tag{6}$$

## 2.2. Converting of objects

The conditions of the existence of objects in the environment are determined by the interaction of physical, chemical, biological and psychical factors. We assume that for each material object from set $N$ there exists such set of environmental conditions in which there is a loss of $\varsigma$ object properties ($\varsigma > 0$) belonging to set $Nt$. If at least one of $\varsigma$ lost features belongs to set $Im$ we are dealing with the transformation of an element from set $N$. In this case the object from set $N$ after the loss of at least one feature from set $Im$ does not belong to set $N$. A set of environment conditions which may influence the transformation of an object from set $N$ is called transformation conditions. Their intensity, at which the object is transformed, is called transformation intensity and is denoted as $\Gamma_t$. Locally occurring environmental conditions of transformation intensity were designated $\gamma_t$. The features of an object from subset $Ct$ have no effect on the variation of transformation intensity of material objects. We assume, however, that the combination of $\xi$ features of set $Vr$ affects the value of the smallest transformation intensity $\gamma_{ti}$ of the $i$-th object from set $N$, according to the formula

$$\gamma_{ti} = f_4\left(vr_1, \ldots, vr_\xi\right).\tag{7}$$

If none of $\varsigma$ lost properties by the $i$-th element belongs to set $Im$ we are dealing with the formation of object from set $N$. After the loss of features not belonging to set $Im$ an object from set $N$ still belongs to set $N$. A set of environment conditions which may influence the formation of an object from set $N$ is called formation conditions. Their intensity, at which the object is formed, is called formation intensity and is denoted as $T_t$. Locally occurring environmental conditions of formation intensity were denoted by $\tau_t$. The features of an object from subset $Ct$ have no effect on the variation of formation intensity of material objects. However, the combination of $\xi$ features of set $Vr$ affects the value of the smallest transformation intensity $\tau_{ti}$ of the $i$-th object from set $N$, according to the formula

$$\tau_{ti} = f_5\left(vr_1, \ldots, vr_\xi\right).\tag{8}$$

Later in this paper the theory of transforming objects is described. The formation of material objects is an important issue because of the possibility of changing their susceptibility to transformation. The theory of random formation of matter is analogous to the described theory.

## 2.3. Types of volume

The main parts of space $V$ are volumes $V_{ai}$. Environmental conditions occurring in them are ~~~e~~~ for the individual objects from set $N$. Thus volume $V_{ai}$ is safe for the $i$-th object from the set $N$. We assume that this volume there is intensive mixing. Its purpose is to homogenise the concentration of objects present in the volume $V_{ai}$.

Volume $V_{ai}$ consists of two parts:

a.   $V_{aci}$ is a part of volume $V_{ai}$, whose subsets are never transformed to other types of volumes,

b.   $V_{ati}$ is a part of volume $V_{ai}$, in which subsets can be transformed to other types of volumes ($V_{\gamma ji}$ and $V_{\beta ji}$).

Between the components of volume $V_{ai}$ for the $i$-th object there are relationships which are described by the following formula

$$V_{ai} = V_{ati} \cup V_{aci} \qquad (9)$$

and

$$V_{ati} \cap V_{aci} = \varnothing. \qquad (10)$$

In the space $V_{ati}$ there are generated transformation volumes $V_{\gamma ji}$. Possible cases of generating transformation volumes are shown in Figure 2. Volume $V_{\gamma ji}$ is the $j$-th transformation volume of the $i$-th material object from set $N$. We assume that transformation volumes generated for the $i$-th element are uniformly dispersed in space $V_{ati}$. Space $V_{ati}$ is incomparably greater than any $j$-th volume $V_{\gamma ji}$, which is described by the formula

$$V_{\gamma ji} \ll V_{ati}. \qquad (11)$$

Volume $V_{\gamma ji}$ is limited from the outside with surface $F_{\gamma aji}$ and from the inside with surface $F_{\gamma \beta ji}$ (Fig. 2). Both surfaces belong to volume $V_{\gamma ji}$ according to the formula

$$F_{\gamma a ji} \in V_{\gamma ji}, \qquad (12)$$

and

$$F_{\gamma \beta ji} \in V_{\gamma ji}. \qquad (13)$$

Component factors of transformation conditions of the $i$-th material object can affect it locally or even pointwise. Component factors of transformation conditions of the $i$-th material object can affect it locally or even pointwise. However, the effects of their action affect the whole object. We assume that the smallest transformation volume of the $i$-th object is equal to its volume at the moment of transformation and amounts $(V_{\gamma ji})_{min}$ (Fig. 3). Occurring in his volume

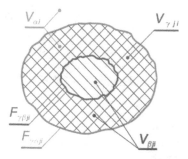

**Figure 2.** Schematic of formed transformation volume $V_{\gamma i}$ mach larger then $(V_{\gamma i})_{min}$: general case.

set of transformation conditions of intensity at least $\gamma_{ti}$ ensures transformation of the $i$-th object from set $N$. Volume $(V_{\gamma ji})_{min}$ is a value characterizing the $i$-th material object from set $N$.

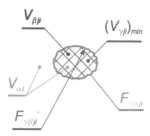

**Figure 3.** Schematic of formed transformation volume $V_{\gamma i} = (V_{\gamma i})_{min}$: special case.

The range of variation of generated in space $V$ transformation volume $V_{\gamma ji}$ by the size of the volume is described by the following formula

$$\left(V_{\gamma ji}\right)_{min} \leq V_{\gamma ji} \leq \left(V_{\gamma ji}\right)_{max}, \tag{14}$$

and by the intensity of transformation conditions by a formula

$$V_{\gamma ji\left(\gamma_t = \gamma_{t min}\right)} \leq V_{\gamma ji\left(\gamma_t\right)} \leq V_{\gamma ji\left(\gamma_t = \gamma_{t max}\right)}. \tag{15}$$

By $\gamma_{tmin}$ and $\gamma_{tmax}$ denoted accordingly minimum and maximum intensity of transformation conditions which can be generated in space $V_{\gamma ji}$.

Transformation volume $V_{\gamma ji}$ is generated at time $t$ $(t > t_0)$ of process transformation duration. It can exist in any time interval $\Delta t > 0$. In this time interval $V_{\gamma ji}$ can increase; if it is greater than

$(V_{\gamma ji})_{min}$ it can decrease or stay unchanged. It can also be displaced randomly to any available place of volume $V_{ati}$. In general form $V_{\gamma ji}$ is described by the relationship in the following formula

$$V_{\gamma ji} = \iint_{D_1} \left[ f_{61ji}(x,y,\gamma,t) - f_{62ji}(x,y,\gamma,t) \right] dxdy - \iint_{D_2} \left[ f_{71ji}(x,y,\gamma,t) - f_{72ji}(x,y,\gamma,t) \right] dxdy \qquad (16)$$

of course at the assumptions described by the relationships

$$f_{61ji}(x,y,\gamma,t) \geq 0 \qquad (17)$$

and

$$f_{71ji}(x,y,\gamma,t) \geq 0. \qquad (18)$$

Flat areas $D_1$ and $D_2$ describe relationships defined by formulas respectively

$$y = f_{63ji}(x,\gamma,t) \qquad (19)$$

and

$$y = f_{73ji}(x,\gamma,t), \qquad (20)$$

the range of changes in $x$ value is described by the relationships defined in the formulas respectively

$$x_{6ji\min} \leq x \leq x_{6ji\max} \qquad (21)$$

and

$$x_{7ji\min} \leq x \leq x_{7ji\max}. \qquad (22)$$

Inside volume $V_{\gamma ji}$ volume $V_{\beta ji}$ is generated. It is not available for non-transformed $i$-th object from set $N$.

Axiom 1.

If linear dimensions of volume $V_{\gamma ji}$ are smaller than doubled linear dimensions of the $i$-th object belonging to set $N$, then the area of $F_{\gamma\beta ji}$ reached by the $i$-th object from the inside does not belong to volume $V_{\alpha i}$ according to

$$F_{\gamma\beta ji}^{\phantom{ji}-} \notin V_{\alpha i}. \tag{23}$$

Since for limit case $(V_{\gamma ji})_{min}$ there is

$$V_{\beta ji} \equiv \left(V_{\gamma ji}\right)_{min}, \tag{24}$$

that is

$$F_{\gamma\beta ji} \equiv F_{\gamma\alpha ji} \tag{25}$$

then

$$F_{\gamma\alpha ji}^{\phantom{ji}-} \notin V_{\alpha i}. \tag{26}$$

and

$$F_{\gamma\alpha ji}^{\phantom{ji}-} \in V_{\beta ji}. \tag{27}$$

Axiom 2.

If the linear dimensions of the volume $V_{\gamma ji}$ are greater than doubled linear dimensions of the $i$-th object belonging to $N$ then area $F_{\gamma\beta ji}$ does not belong to $V_{\alpha i}$ volume according to the formula

$$F_{\gamma\beta ji} \notin V_{\alpha i}. \tag{28}$$

For the case for which occurs a relationship described by a formula (28), volumes closed by area $F_{\gamma\beta ji}$ do not belong to space $V_{\alpha i}$.

Volume $V_{\beta ji}$ is closed by external surface $F_{\gamma\alpha ji}$.

Axiom 3.

If the linear dimensions of the volume $V_{\gamma j i}$ are smaller than doubled linear dimensions of the $i$-th object belonging to $N$ then area $F_{\gamma \beta j i}$ reached by the $i$-th object from the inside belongs to $V_{\beta j i}$ volume according to the formula

$$F_{\gamma \beta j i}^{-} \in V_{\beta j i}. \tag{29}$$

Axiom 4.

If the linear dimensions of the volume $V_{\gamma j i}$ are greater than doubled linear dimensions of the $i$-th object belonging to $N$ then area $F_{\gamma \beta j i}$ does not belong to $V_{\beta j i}$ according to the formula

$$F_{\gamma \beta j i} \in V_{\beta j i}. \tag{30}$$

For the case for which occurs relationship described by the formula

$$V_{\gamma j i} > \left( V_{\gamma j i} \right)_{\min}, \tag{31}$$

volumes closed by area $F_{\gamma \alpha j i}$ do not belong to $V_{\beta j i}$ volume according to formula

$$F_{\gamma \alpha j i} \notin V_{\beta j i}. \tag{32}$$

If the linear dimensions of the volume $V_{\gamma j i}$ are equal to the doubled linear dimensions of the $i$-th object belonging to $N$ then area $F_{\gamma \beta j i}$ may be replaced by a segment (for example for cylindrical surfaces) or even a point (for example for spherical surfaces). They will belong to volume $V_{\beta j i}$ and they will not belong to volume $V_{\alpha i}$.

Volume component of the space $V$, apart from the above mentioned kinds of volumes, is $V_{\delta i}$. This volume meets the following conditions:

1. is safe for the $i$-th object from set $N$,

2. there is not mixing in it,

3. there is a possibility of object migration from volume $V_{\alpha i}$ to $V_{\delta i}$ and vice versa,

4. subsets of this volume are not transformed to other volumes.

We assume that there is a relationship

$$V_{\delta i} < V_{\alpha i}. \tag{33}$$

Volumes $V_{\delta i}$ may be dispersed in space $V$ as volumes $V_{\gamma j i}$.

## 2.4. Transformation process

We assume that the above described events: introduction, removal, multiplication, and exclusion are not the events of the investigated process of random transformation of dispersed material objects.

In a given material medium $V$ occurs process of random transformation of objects belonging to set $N$. It runs as follows:

At any moment $t$ $(t > 0)$ for the $i$-th object in space $V_{ati}$ there are generated randomly $p$ transformation volumes $V_{\gamma ji}$, where $p$ is a natural number. The $i$-th object belonging to set $N$ is in the space safe volume $V_{ai}$ described by a relationship

$$V_{\alpha i} = \bigcap_{j=1}^{p} V_{\alpha ji}. \tag{34}$$

We assume that random transformation of objects is independent events.

It is assumed that during transformation process the number of produced volumes $V_{\gamma ji}$ is large and may change over time, according to the formula

$$p = f_8(t). \tag{35}$$

Due to relative displacement the $i$-th material object is introduced for time $t_{ti}$ to appropriate volume for its transformation $V_{\gamma ji}$. This is done by the surface $F_{\gamma aji}$ which limits volume $V_{\gamma ji}$ (Fig. 2 and 3). The considered element of set $N$ remains unconverted if at least its one point is beyond volume $V_{\gamma ji}$. Transformation of the $i$-th object occurs simultaneously with its complete introduction to transformation volume $V_{\gamma ji}$. According to the conditions given in Section 2.2 transformation occurs when at least one of the $\varsigma$ features belonging to set $Ba$ is lost. Only transformed in volume $V_{\gamma ji}$ $i$-th object may be dislocated to volume $V_{\beta ji}$.

Theorem about transformation of dispersed matter:

For each material object $n_i$ from set $N$ there exist such local conditions of transformation $\gamma_{ti}$ belonging to set $\Gamma_t$, that if there is time $t_{ti}$, in which object $n_i$ is included in volume $V_{\gamma ji}$ of transformation properties $\gamma_{ti}$ then starting from time $t_{ti}$ object $n_i$ does not belong to set $N$.

This theorem recorded by means of quantifiers has the following form:

$$\left(\forall n_i \in N\right)\left(\exists \gamma_{ti} \in \Gamma_t\right)\left(\left(\exists t_t\right)n_i \subset V_{\gamma ji} \Rightarrow \left(\forall t \geq t_t\right)n_i \notin N\right), \tag{36}$$

and its falsification is included in the record

$$\left(\forall n_i \in N\right)\left(\exists \gamma_{ti} \in \Gamma_t\right)\left(\left(\exists t_t\right)n_i \subset V_{\gamma ji} \Rightarrow \left(\forall t \geq t_t\right)n_i \in N\right). \tag{37}$$

Proof of the theorem

The properties of an object from set $N$ belonging to set $Ba$ (Fig.1) are marked by $ba$. The relationship is introduced

$$Z_1(n;ba) \tag{38}$$

which expresses the statement that object $n$ has the property $ba$. We can record the following observations:

object $n_i$ has all the properties ba from set $Ba$

$$\left(\forall ba \in Ba\right)\left(Z_1\left(n_i;ba\right)\right), \tag{39}$$

and there is a property $ba$, which object $n_i$ does not have

$$\left(\exists ba \in Ba\right)\left(\neg Z_1\left(ba;n_i\right)\right). \tag{40}$$

Let $N$ be the set of all elements fulfilling condition $W(n)$ described by a logical statement (39)

$$\{n_i \in N : ((\forall\, ba \in Ba)(Z_1(n_i;ba)))\}. \tag{41}$$

The relationship is introduced

$$Z_2(V_\gamma;\gamma_t) \tag{42}$$

expressing the statement that volume $V_\gamma$ has the property $\gamma_t$ of object transformation $n$. We can write the following observation:

object $V_{\gamma ji}$ has got a property $\gamma_{ti}$ of object transformation $n_i$

$$\left(\exists\gamma_{ti} \in \Gamma_t\right)\left(Z_2\left(V_{\gamma ji};\gamma_{ti}\right)\right). \tag{43}$$

Let $P$ be the set of all elements fulfilling the condition $W_2(V_{\gamma ji})$ described by a logical statement (39)

$$\{V_{\gamma ji} \in P : ((\forall\, ba \in Ba)(Z_1(n_i;ba)))\}. \tag{44}$$

Let $t_{ti}$ denotes time in which object $n_i$ was introduced to volume $V_{\gamma ji}$. From the moment $t_{ti}$ is true the statement about object $n_i$

$$\left(\forall t \geq t_{ti}\right)\left(\exists ba \in Ba\right)\left(\neg Z_1\left(ba; n_i\right)\right). \tag{45}$$

It is contrary to the assumption (31) so that from the moment $t_{ti}$ the object $n_i$ cannot be the element of set $N$ according to

$$n_i \notin N \ \blacksquare. \tag{46}$$

Theorem about creating families of transformation volumes

If at any point 2 belonging to space $V_{ati}$ environmental conditions of the transformation would function $\gamma_{t2}$ and in set $N$ there would be $r$ objects of the transformation conditions $\gamma_{ti}$ fulfilling the relations

$$\gamma_{t2} \geq \gamma_{ti} \tag{47}$$

Then in point 2 there will be generated family consisting of $r$ transformation volumes for $r$ objects from set $N$.

Conclusions:

1.  If set $N$ consists of $n$ elements and conditions of transformation $\gamma_{ti}$ of any element from set $N$ satisfy the relation

$$\gamma_{tmin} \leq \gamma_{ti} \leq \gamma_{tmax} \tag{48}$$

then the family can include from 1 to $n$ transformation volumes.

2.  If in any point 2 belonging to space $V_{ati}$ would function conditions of transformation $\gamma_{t2}$ fulfilling the relationship

$$\gamma_{t2} < \gamma_{tmax} \tag{49}$$

and in set $N$ there would exist $s$ objects of transformation conditions greater then $\gamma_{t2}$ then in space V there are s objects from set $N$ which are not subjected to transformation in point 2.

Volume $V_{yji}$ is a transformation volume of the $i$-th material object which belongs to $j$-th family.


# 3. General phenomenological model

Figure 4 shows the general set of possibilities for generating volumes associated with the transformation of objects from set $N$ distributed in space $V(t)$. Each vertical segment with the opposite ends located on segments AB and CD denotes space $V(t)$. On each subsequent vertical section there are marked up divisions of volume $V$ in result of generated one transformation volume. The whole rectangle ABCD includes a set of all possible divisions of space $V$ according to one defined system.

For any $i$-th object from set $N$ rectangle ABCD is divided into five parts with four segments. The first part, contained between segments LM (marked with blue dashed line) and IJ (marked with red dashed line) is a set of possible volumes $V_{\alpha ji}$. From the bottom a single volume $V_{\alpha i}$ is limited by surface $F_{\gamma\beta ji}$. Depending on the size of generated volume $V_{\gamma ji}$ this surface according to formula (23) conditionally does not belong or according to formula (28) at all does not belong to volume $V_{\alpha i}$. Space $V_{\alpha i}$ is safe for both the $i$-th object and other objects from set $N$ of the smallest transformation volume not greater than that which is appropriate for the $i$-th object. In volume $V_{\alpha ji}$ takes place intensive mixing. Its purpose is to homogenize the dispersion of objects present in it.

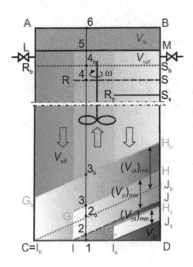

**Figure 4.** General set of space divisions $V(t)$ resulting from generating volume $V_{\gamma ji}$ for one hypothetical generating system.

In Figure 4 the area included between segments GH (marked with orange dashed line) and IJ (marked with red dashed line) is a set transformation volumes that can be generated. Single volume $V_{\gamma ji}$ is limited by surfaces: $F_{\gamma\alpha ji}$ from the top and $F_{\gamma\beta ji}$ from the bottom. Both these surfaces belong to volume $V_{\gamma ji}$, which is the result of assumptions accepted in formulas (12) and (13). In volume $V_{\gamma ji}$ occurs a transformation if the $i$-th object and other objects from set $N$ of minimal transformation volume in accordance with that which is appropriate for the $i$-th object i.e. $(V_{\gamma i})_{min}$. During the process at time $t$ there are $p$ transformations volumes of the $i$-th object according to relationship (34) are they are uniformly dispersed in space $V_{\alpha ti}$. The size of the volume $V_{\gamma ji}$ in Figure 4 is defined by two parameters: $(V_{\gamma ji})_{min}$ and red colour intensity. The parameter – red colour intensity is a visualization of functional dependence.

Another area in Figure 4, contained between segments GH (marked with orange dashed line) and CD (marked with brown solid line) is a set of volumes $V_{\beta ji}$. Such single volume is limited by surface $F_{\gamma aji}$. Depending on the size of generated volume $V_{\gamma ji}$ this surface according to (27) conditionally belongs to $V_{\beta ji}$ or according to formula (32) does not belong to $V_{\beta ji}$. During the process at time $t$ volume $V_{\beta ji}$ is no more than $p$. They are linked to appropriate volumes $V_{\gamma ji}$ and dispersed in space $V_{ati}$. In volume $V_{\beta ji}$ the existence of unconverted objects from set $N$ is impossible, for which minimal transformation volume is not smaller than minimal transformation volume appropriate for the $i$-th object. The size of the volume $V_{\beta ji}$ in Figure 4 is defined by two parameters: vertical segment with ends located on segments GH and CD and brown colour intensity. The consequence of increasing volume $V_{\gamma ji}$ is increased volume $V_{\beta ji}$ and decreased volume $V_{ati}$. The size of volume $V_{ati}$ in Figure 4 is defined by two parameters: vertical segment with ends located on segments RS and IJ and blue colour intensity.

Between segment AB and segment LM there is volume $V_{\delta i}$.

Division of vertical segment 16 reflects the division of space $V$ which appears after generation of the $j$-th volume for the $i$-th object $n_i$:

- segment $\langle 1, 3)$ corresponds to volume $V_{\beta ji}$,

- segment $\langle 2, 3\rangle$ corresponds to volume $V_{\gamma ji}$,

- segment $(2, 4\rangle$ corresponds to volume $V_{atji}$,

- segment $(4, 5\rangle$ corresponds to volume $V_{acji}$,

- segment $(5, 6\rangle$ corresponds to volume $V_{\delta i}$.

In Figure 4, as for the $i$-th object, it is possible to determine components of space volume $V$ for the objects of the highest and smallest transformation volume marked $(V_{\gamma b})_{min}$ and $(V_{\gamma s})_{min}$ respectively.

Between segments AC and BD it is possible to lead infinite number of vertical lines. Their division into sections, presented above, describes the field of possibilities of generation, in space $V$, the transformation volume $V_{\gamma ji}$ and connected with them volumes $V_{\beta ji}$. For segment AC transformation volumes are generated according to Figure 3. The divisions are on the right side of Figure 4 together with the segment BD is the general case shown in Figure 2.

At any moment of the transformation process in volume $V$ a finite number transformation volume is generated.

Phenomenological model of random transformation of dispersed matter is constructed as follows:

1.   The possibilities of generating transformation volumes in space $V$ are shown on the system map appropriate for the given transformation process which is analogous to that presented in Figure 4. In space $V$ transformation volumes according to many system maps can be generated simultaneously.

2.  Phenomenological model consists of $p$ layers formed by $p$ transformation volumes generated at a given moment in volume $V$ according to system maps.

3.  The model is constructed in such a way that the sum of sets of all volumes is equal to $V$. The sum of all volumes safe for the $i$-th objects located above the generated in a given moment its transformation volume is equal to $V_{ai}$, according to the equation (33).

4.  Material objects belonging to set $N$, in space $V(t)$, were reduced to a point. One of their features is volume $V_{ni}$. On system maps (Fig. 4) and in the phenomenological model the size of volume is defined by vertical segments, so the $i$-th object will be a vertical segment of the length corresponding to volume $(V_{\gamma ji})_{min}$. This object can be displaced in space $V_{ai}$ and $V_{\delta i}$ till the transformation moment in volume $V_{\gamma ji}$.

## 4. General mathematical model

In general, the concentration of elements of set $N$ in volume $V(t)$ is determined by the number of not transformed objects $n$ per volume $V$ according to the formula

$$S(t) = \frac{n(t)}{V(t)}. \tag{50}$$

At the initial moment $t_0 = 0$ it will amount

$$S_0 = \frac{n_0}{V_0}. \tag{51}$$

We assume that for the process duration moment $t_0 = 0$ in space $V$ volumes $V_{\gamma ji}$, $V_{\beta ji}$, $V_{ai}$ and $V_{\delta i}$ are generated. Not transformed $i$-th object can occur only in the appropriate volumes $V_{ai}$ and $V_{\delta i}$. From formulas (1) and (7) results the relationship between the volumes for the individual objects from set $N$ (Fig. 4) given in the formula

$$V_{ab} \subset \dots \subset V_{ai} \subset \dots \subset V_{as}. \tag{52}$$

For relationship (52) equivalence

$$\left(V_{\gamma s}\right)_{min} = \left(V_{\gamma b}\right)_{min} \Leftrightarrow V_{as} = V_{ab} \tag{53}$$

or

$$\left(V_{\gamma s}\right)_{min} \left\langle \left(V_{\gamma b}\right)_{min} \Leftrightarrow V_{ab} \langle V_{as} \right. \tag{54}$$

can be true. In the case included in formula (53) in the whole volume $V_a$ defined by formula

$$V_\alpha = \bigcup_{i=1}^{n}\bigcap_{j=1}^{p} V_{\alpha ji} \tag{55}$$

occurs uniform dispersion of the elements of homogeneous set $N$. However, from the given in formula (54) alternative equivalence result the uneven dispersion of inhomogeneous objects from set $N$ in volume $V_a$. Each object $n_i$ characterized by feature $vr_i$ which distinguishes it from other objects from set $N$ results in introduction of the additional area $F_{\gamma\beta i}$ (Fig. 2 and 3). Each additional area $F_{\gamma\beta i}$ introduced in space $V$ divides volume set $V_a$ (Fig. 4). The resulting parts can differ with the concentration of the objects contained in them. The process of ideal mixing ensures homogeneity of the dispersion only within a volume limited by neighboring areas e.g. $F_{\gamma\beta(i-1)}$ and $F_{\gamma\beta i}$, $F_{\gamma\beta i}$ and $F_{\gamma\beta(i+1)}$. In the case covered by formula (54) due to the relationships given in formulas (5) and (11) as well as uniform dispersion of volume $V_\gamma$ in space $V_a$ we can use average concentration of dispersed material objects. Thus, after starting the process, to the generated volumes $V_{\gamma ji}$ there will be introduced randomly appropriate objects $n_i$ of set $N$ dispersed in volume $V_a$. Elements from set $N$ being in volumes $V_{\delta i}$ do not participate directly in the transformation process. In moment $t_0=0$ of process duration the number of unconverted objects $n(t)$ in volume $V_a$ is defined by the equation

$$n_{\alpha 0} = n_0 - n_{\delta 0} \tag{56}$$

Average concentration of untransformed objects in volumes occupied by them volume $V_a$ is described by formula

$$\left(\overline{S_\alpha}\right)_{t_0=0} = \frac{n_{0\alpha}}{V_\alpha} \tag{57}$$

where $n_{\delta 0}$ denotes initial number of elements from set $N$ in volumes $V_{\delta i}$. Number of objects which were transformed $n_{d0}$ for time $t_0$ is defined by the formula

$$n_{d0} = 0 \tag{58}$$

The degree of objects transformation $X(t)$ determined by the quotient of the number of transformed objects $n_d$ to the initial number of unconverted objects $n_0$ for $t = 0$ equals 0, according to the formula

$$X = \frac{n_{d0}}{n_0} = 0. \tag{59}$$

At any moment of process duration $t$ the number of transformed objects amounts $n_d(t)$. Converted objects may be located in any place of volume $V$. The number of unconverted objects, being exclusively in volume $V_\alpha$, is defined by the difference of the initial number of objects $n_{\alpha 0}$ and converted objects $n_d$ after time $t$ of process duration corrected by a number of objects transferred between volumes $V_{\alpha i}$ and $V_{\delta i}$ and the number of objects described by the formula (6), which is described by equality

$$n_\alpha = n_{0\alpha} - n_d + n_{\delta\alpha} - n_{\alpha\delta} + \Delta n_A, \tag{60}$$

where:$n_{\delta\alpha}$ – means number elements of set $N$ transferred from volume $V_{\delta i}$ to $V_{\alpha i}$,

$n_{\alpha\delta}$ – means number of elements of set $N$ transferred from volume $V_{\alpha i}$ do $V_{\delta i}$.

After time $t$ of process duration, to transformation volumes $V_{\gamma ji}$ there are introduced untransformed objects of average concentration $\left(\overline{S_\alpha}\right)_t$ defined by the number of unconverted $n_\alpha$, per volume $V_\alpha$. This is shown in formula

$$\left(\overline{S_\alpha}\right)_t = \frac{n_\alpha}{V_\alpha}. \tag{61}$$

Of course general concentration of unconverted objects in volume $V$ will be defined by the relationship

$$S = \frac{n}{V}, \tag{62}$$

where

$$n = n_\alpha + n_\delta. \tag{63}$$

It will be shown, for example, after stopping the process. The degree of transformation of objects after time $t$ of the transformation process is described by the formula

$$X = \frac{n_d}{n_0 + \Delta n_A}. \tag{64}$$

The increase of transformed objects $dn_d$ in all transformation volumes $V_{\gamma ji}$ after the lapse of any small time interval $dt$ is defined by the formula

$$dn_d = \left(\overline{S_\alpha}\right)_t dV.$$

(65)

The external area limiting the generated for the $i$-th object the $j$-th volume $V_{\gamma ji}$, consists of active and inactive part. Through the active part there can be introduced or removed the $i$-th element of set $N$ while the inactive part is not available for such transfer. The active part of the surface is divided into surface $F_{\gamma\alpha\rightarrow}$, through which the $i$-th object can be introduced to $V_{\gamma ji}$ and surface $F_{\gamma\alpha\leftarrow}$, through which the transformed or not transformed $i$-th object can be introduced from $V_{\gamma ji}$ according to the formula

$$F_{\gamma\alpha ji} = F_{\gamma\alpha ji\rightarrow} + F_{\gamma\alpha ji\leftarrow}.$$

(66)

The sum of surfaces limiting the transformation volumes $V_{\gamma ji}$ generated in space $V$ for the $i$-th object is described by the formula

$$F_{\gamma\alpha i} = \sum_{j=1}^{p} F_{\gamma\alpha ji}.$$

(67)

The sum of surfaces limiting the volumes $V_{\gamma ji}$ generated in space $V$ for all objects form set $N$ is described by the formula

$$F_{\gamma\alpha} = \sum_{i=1}^{n}\sum_{j=1}^{p} F_{\gamma\alpha ji}.$$

(68)

Volume $dV$ displaced from space $V_{\alpha ji}$ to volume $V_{\gamma ji}$ in time increase $dt$ depends on the size of limit area $F_{\gamma\alpha\rightarrow}$, through which $dV$ is displaced from volume $V_{\alpha ji}$ to volume $V_{\gamma ji}$ and on the average speed of its displacement $u$, according to the formula

$$dV = uF_{\gamma\alpha\rightarrow}dt.$$

(69)

After substituting equations (61) and (69) to equation (65) we get the equation

$$dn_d = k\left(n_\alpha\right)dt$$

(70)

describing the increase of objects transformed in generated volumes $V_{\gamma ji}$. Process rate constant $k$ is described by the relationship

$$k = \frac{F_{\gamma \alpha \rightarrow}}{V_\alpha} u.$$

(71)

According to formulas (16), (19) and (20) parameters $F_{\gamma \alpha \rightarrow}$ and $V_\alpha$ depend on time $t$. Dislocation rate $u$ can also be a function of process duration

$$u = f_9(t).$$

(72)

On the basis of formula (70) the loss of unconverted objects in set $N$ can be expressed by the equation

$$dn = -k\left(n_\alpha\right)dt.$$

(73)

After substituting (60) to the relationship (73) we get

$$dn = -k\left(n_{0\alpha} - n_d + n_{\delta\alpha} - n_{\alpha\delta} + \Delta n_A\right)dt.$$

(74)

The total balance of the loss of objects $dn_d$ in time interval $dt$ will be equal on the right side of the formula (74) with the opposite sign

$$dn_d = k\left(n_0 - n_d + n_{\delta\alpha} + n_{\alpha\delta} + \Delta n_A\right)dt.$$

(75)

In case

$$\Delta n_A = 0$$

(76)

the relationship (76) will be simplified to

$$dn_d = k\left(n_0 - n_d\right)dt.$$

(77)

After separation of variables in equation (77) and integration of both sides we get the relationship

$$\ln\left(\frac{n_0}{n_0 - n_d}\right) = kt, \tag{78}$$

often used to describe the kinetics of the course of transformation of various material objects. In case of homogeneity of set $N$ ($Vr$ is an empty subset) relationship (78) is a final description of the transformation process. Let's consider the case when the object of set $N$ differ with only one feature $vr_1$, which has a considerable effect on the course of the process. In order to analyse it, the whole range of changes should be divided into $m$ such intervals in which changes in features have no significant effect on the course of the process. Then the course of transforming objects from any interval $\zeta$ - th such that

$$\zeta \in \langle 1, m \rangle, \tag{79}$$

describes the relationship

$$\frac{n_{0\zeta}}{n_0} dn_{d\zeta} = \frac{n_{0\zeta}}{n_0} k_\zeta \left(n_{0\zeta} - n_{d\zeta}\right) dt \tag{80}$$

taking into account the existence of a whole set $N$.

Description of the transformation of the entire set of objects including the division of set $N$ into intervals is described by the relationship

$$\sum_{\zeta=1}^{m} \frac{n_{0\zeta}}{n_0} dn_{d\zeta} = \sum_{\zeta=1}^{m} \frac{n_{0\zeta}}{n_0} k_\zeta \left(n_{0\zeta} - n_{d\zeta}\right) dt. \tag{81}$$

After the separation of variables (80) and integration of both sides we get

$$\sum_{\zeta=1}^{m} \frac{n_{0\zeta}}{n_0} \ln \frac{n_{0\zeta}}{n_{0\zeta} - n_{d\zeta}} = \sum_{\zeta=1}^{m} \frac{n_{0\zeta}}{n_0} k_\zeta t. \tag{82}$$

Relationship between speed constants in equations (77) and (80) is defined by a formula

$$\Phi \sum_{\zeta=1}^{m} \frac{n_{0\zeta}}{n_0} k_\zeta = k. \tag{83}$$

Coefficient $\Phi$ describes the relationship

$$\Phi = \ln\frac{n_0}{n_0 - n_d}\left(\sum_{\zeta=1}^{m}\frac{n_{0\zeta}}{n_0}\ln\frac{n_{0\zeta}}{n_{0\zeta} - n_{d\zeta}}\right)^{-1}.$$ (84)

## 5. Examples of application

The technological process based on the theory of random transformation of dispersed matter is the disintegration of microbial cells. At present many compounds coming from the inside of the microorganism cells have commercial application. They are used among others in the food, pharmaceutical, cosmetic, chemical industry as well as medicine and agriculture. In order to isolate the desired compounds it is usually necessary to destroy the cell walls and cytoplasmic membranes. The process of disintegration of microorganisms is carried out by different methods: physical, chemical and biological.

The technical means used to implement the process on an industrial scale are bead mills and high pressure homogenizers. On a laboratory scale there are often used vibrating mixers, ultrasonic homogenizers and enzymatic methods. All the listed methods of the process involve the random effects of the factor which destroys the cell walls of microorganisms dispersed in the liquid. The difference between them depends mainly on the method of generating the transformation volume of dispersed matter.

During the disintegration of microorganisms the suspension occupies volume $V$. It is constant in time. The process is carried out usually for the optimal initial concentration of biomass. The initial number of microorganisms is determined $n_0$. We consider the case of batch operation (constant charge). During the process microorganism cells are not added from the outside (we assume the sterility of conditions of disintegration) nor are they removed outside. The process duration compared to the lifetime of microorganisms and the time needed to form new cells is very short. So it can be assumed that the change in the number of objects described by the formula (6) satisfies the relationship (76).

In the case of disintegration of microorganisms in of bead mills the transformation consists in the disruption of microbial cell walls. Destruction volumes $V_{\gamma ji}$ are generated by circulating filling beads. To carry out the process in the bead mill, high degree of filling the working chamber with beads and high rotational speed of the agitator are applied. It can be assumed that at any time during the process for fixed working conditions number $p$ (Eq. (35)) is constant in time and its value is high. Thus the relations given in the formulas

$$\sum_{j=1}^{p}V_{\gamma ji} = const,$$ (85)

$$\sum_{j=1}^{p}V_{\beta ji} = const$$ (86)

are fulfilled.

General diagram of cell disruption between the spherical surfaces is shown in Figure 5. It concerns the range of the distribution of suspension volume $V$ to $V_{aji}$, $V_{\gamma ji}$ and $V_{\beta ji}$ shown on the right in Figure 4 in the area adjacent to the segment BD. Volumes $V_{ac}$ occur close to the inside surface of the mill chamber in the case when filler elements have a diameter substantially greater than the dimensions of the cells of microorganisms. In a bead mill volumes $V_{ac}$ occur at all surfaces of the working chamber and agitator. They are distant by a distance similar to the size of the largest cells and the thickness of their layer is slightly smaller than the radius of the smallest filling beads. In these volumes cells are never disrupted.

For a properly constructed mill chamber volumes $V_\delta$ (may be slots at the interface between two structural elements) are negligibly small and insignificant in terms of technology, especially when conducting sterilization of equipment between the processes. In the analyzed case it was assumed that disintegrated microorganisms have an ellipsoidal shape. After the limit deformation of the $i$-th cell, its walls are disrupted. (Fig. 5). The generated transformation volume $V_{\gamma ji}$ is limited by surfaces: active $F_{\gamma aji}$ (orange dashed line), $F_{\gamma \beta ji}$ (red dashed line) and two inactive spherical of filling elements. Limiting surfaces $F_{\gamma aji}$ and $F_{\gamma \beta ji}$ belong to volume $V_{\gamma ji}$. Its axis of symmetry is axis OO. To the presented in Figure 5 volume $V_{ai}$, in which $i$-th living cell can be present, does not belong the volume limited by two spherical surfaces and surface $F_{\gamma \beta ji}$. Straight line OO is a symmetry axis of this volume. Surfaces limiting volume $V_{aji}$ do not belong to it. The volume unavailable for the living $i$-th cell $V_{\beta ji}$ is limited by two spherical surfaces and surface $F_{\gamma aji}$. Limiting surfaces do not belong to $V_{\beta ji}$. The axis of symmetry of volume $V_{\beta ji}$ is also straight line OO. In the special case the line dividing volume $V$ in an AC position (Fig. 4), there is generated volume $(V_{\gamma ji})_{min}$ shown in Figure 6. It is limited by two spherical surfaces and active surface $F_{\gamma aji}$ (orange dashed line). Volume symmetry axis $(V_{\gamma ji})_{min}$ passes through points O and O. Surface $F_{\gamma aji}$ does not belong to generated volume $V_{\beta ji}$. If all points of the $i$-th object are not introduced to $(V_{\gamma ji})_{min}$ it will not be transformed.

For monogenic set $N$ ($Vr$ is an empty subset) cell disintegration process, taking into account the above assumptions, is described by differential equation (77). The case when the objects of set $N$ differ with only one feature $vr_1$ that has a significant impact on the course of cell disruption is included in equation (80) and (81). The process of release of intracellular compounds, using the theory of transformation of dispersed matter, has been widely described by Heim at all [12], Heim and Solecki [13] as well as Solecki [10 and 14]. These works included more complex cases of the course of the process, causes of nonlinearity of kinetics were given, and the effect of concentration of microorganism suspension was explained as well as disappearance of the largest size fraction during the process for very low concentrations of the suspension.

There are many devices implementing the process of disintegration of microorganisms due to critical stresses in the cell walls caused by stress in the liquid. Best known are: high pressure homogenizers [6, 15-18], French press [19], Ribie press [20], Chaikoff press [21]. The technical method of process realization in the above mentioned equipment is similar and consists in disruption of microorganisms during pumping suspension through a valve under high pressure. High performance in continuous operation and fairly wide range of diversity of disintegrated microorganisms made that high pressure homogenizers are widely used on a

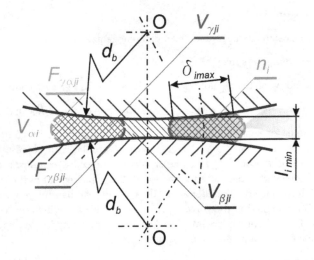

**Figure 5.** Model of cell disruption during non-axial hitting with spherical elements – general case [10].

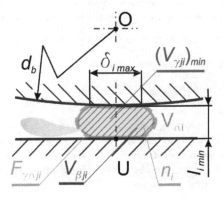

**Figure 6.** Model of cell disruption during axial hitting with spherical element and plane – special case [14].

technical scale. The general model of disintegration of mechanisms, developed by Hetherington et al. [6], and then modified by Sauer et al. [15] is included in a formula

$$\ln\left(\frac{R_m}{R_m - R}\right) = k N^{b_2} P^{b_1},$$ (87)

where: $R$ concentration of released proteins [mg/g],

$R_m$ - maximum concentration of released proteins; [mg/g],

$k$ - process rate constant; [1/s],

$N$ - number of suspension passage cycles through homogenizer; [-],

$P$ - suspension pumping pressure, [MPa].

$b_1$ - exponent depending on the type of microorganisms and their growth conditions; [-],

$b_2$ - exponent including the effect of suspension concentration on the course of the process; [-].

Pumping pressure of the suspension of microorganisms is within the range from 50 to 120 MPa. Transformation volumes $V_{\gamma ji}$ are generated, according to the results of research conducted by Keshavarz Moore at al. [16] and Engler [17] on the cell disruption mechanism, in homogenization zone within the valve unit and with the impingement in the exit zone. The results of research conducted by Lander at al. [18] showed that disruption of cells is mainly due to shearing of the liquid in the valve unit and as a result of cavitation occurring in the impingement section, where the stream of suspension hits the impact ring and follows an implosion of bubbles caused by the increased pressure.

High-frequency ultrasounds (in the supersonic wavelength range 15 - 25 kHz) are used for disruption of dispersed in a liquid microbial cells and releasing contained in them intracellular compounds [22-25]. The mechanism of microorganisms disintegration is associated with the occurrence of cavitation induced by ultrasound and hypothetically runs as follows:

1. Passing sound wave causes the thickening and thinning of the liquid.

2. During the thinning of the liquid occurs nucleation and growth of gas and vapor bubbles.

3. During the thickening of the liquid bubble implosion occurs at a rate not less than the speed of sound (hence the loud roar accompanying cavitation).

4. Bubbles which are not adhering to the cells of microorganisms are sinking evenly in all directions. Bubbles adhering to cells are sinking from the free side so that the surface of the liquid with a powerful force strikes the cell wall breaking it and releasing intracellular compounds. The striking force may be so powerful that the released compounds are often destroyed and free radicals are formed.

Daulah [22], using the theory of local isotropic turbulence Kolmogorov [26 and 27], presented a description of the process of ultrasonic disintegration of cells of baker's yeast as a model

$$1 - S_p = \exp(k \cdot t), \tag{88}$$

where: $S_p$ – concentration of released proteins, [g/kg]

$t$ – duration of the process; [min],

$k$ – process rate constant; [1/min].

Dependence of constant $k$ on energy dissipation $P_d$ is described by the relationship

$$k = \zeta \cdot \left(P_d - P_c\right)^{0,9} \tag{89}$$

where: $\zeta$ - constant; [kg/J],

$P_c$ - threshold energy dissipation ensuring occurrence of cavitation; [J/kgs].

Confirming the above results experimentally for brewing yeast showed no effect of suspension concentration on constant k and its proportionality to the energy dissipation level [25].

The process of disinfection consists in an impact of physical or chemical agents in a limited gas, liquid or solid medium on biological contaminants. They may be it viruses, bacteria and their spores, protozoa and their cysts and eggs of parasites. The purpose of disinfection is to destroy the above mentioned objects and to prevent in a required, limited period of time their re-growth. The disinfectants selected according to pathogens cause: an irreversible destruction of cells, disruption of metabolic processes, disruption of biosynthesis and growth. An example of such process is chemical disinfection of water [28]. Its kinetics is often described by Chick's model [3] in form of equation (77). The dependence of the rate constant

$$k = AC^s t \tag{90}$$

taking into account the disinfecting power coefficient $A$ and the concentration of disinfectant $C$ was given by Watson [4]. Exponent $s$ is dependent on the type of disinfectant and medium's pH.

The developed theory can be used to model the impact of population of predator-prey or competition type. If we assume that:

- change objects from the set of $N$ ($f_{31}$ from Eq. (6)) in equation (75) is described under the law of Malthus by the formula

$$f_{31} = a_1 N \tag{91}$$

- rate constant from equation (75) will be marked as $k_\gamma$ and is written in the form

$$k_\gamma = a_2 P, \tag{92}$$

directly dependent on the generated families of transformation volume (see formula (71)),

- unconverted objects $N$ cause an increase in the families of transformation volume of families in the form of $a_3 NP$,

in the absence of objects $N$ ($N$ is an empty set), there is an exponential decrease in the number of families of transformation volume $- a_4 P$, the received equations in the form

$$dN = N\left(a_1 - a_2 P\right)dt \tag{93}$$

and

$$dP = P\left(a_3 N - a_4\right)dt \tag{94}$$

are a model for predator-prey for the Lotka-Volterra system [1 and 2]. The coefficients $a_1, a_2, a_3$ and $a_4$ are constants with positive values.

On the basis of the presented theory it is possible to build other more realistic models of predator-prey system taking into account e.g. natural selection, nutrient profile, the effect of age on the activity of predators and many other factors. It is also possible to model chemical reactions, and assuming the generation in space $V$ of different types of transformation volume, including reversible transformation, enzymatic reactions can be modeled [5].

The theory can be used to study and model the action of immune system [10]. It can also be employed in constructing artificial immunology and controlling the support of various therapies. Simple applications cover responses of the immune system to viral, bacterial, fungal and parasitic infections. The immune response depends on the type, properties, portal of entry and severity of infection and the state of organism. Application of the theory is illustrated by viral infection. Virus replication in the host cells in space $V$ is represented by the function in Eq. (6). The aim of an immune response is to inhibit virus replication in the cell and its spread to other cells. Next goals include the elimination of transformed objects from the organism and development of long-lasting immunity. Initially, transformation volumes $V_{\gamma ji}$ are formed as a result of action of the complement system, interferon (IFN) and natural killer cells (NK) within non-specific anti-viral response. At the next stage, neutralizing antibodies, mainly of class IgG (immunoglobulin G), prevent infection of other cells. Their fragment Fab (fragment antygen binding) binds to antigens of the virus, while fragment Fc (fragment crystallizable) to relevant receptors on NK cells, macrophages and others. This enables phagocytosis and cellular cytotoxicity dependent on antibodies, but probably also immediate destruction with the use of the complement system. In the case of infections through mucous membranes of the intestinal tract and respiratory system the same role is played by IgA (immunoglobulin A) antibodies. In development of humoral immune response the active role is played by CD4$^+$ lymphocytes. In this case antigen proteolysis occurs in endosome. Cytotoxic CD8$^+$ lymphocytes play the crucial role in the response to viral infections. They function inside the cells. They recognize a viral antigen on cells transformed by the infection in association with MHC (major histocompatibility complex) class I molecules. In the infected cells, cytotoxic lymphocytes can induce a synthesis of nucleases destroying genetic material of the virus and enhance IFN synthesis. Responses of the immune system occurring at subsequent stages of antigen destruction according to different mechanisms can be described by the relation analogous to Eq. (81).

Many pathogens developed such properties which allow them to avoid non-specific and specific immune response. High variability of a pathogen caused by differences in the gene sequence region leads to a delay of specific immune response. This is so in the case of influenza, hepatitis C and HIV (human immunodeficiency virus) viruses. Ho et al. [9] studied the effects of HIV treatment with inhibitors. A linear model based on experimental data in the form concordant with Eq. (77) well described the healing process at the first stage of the therapy. At the final stage, however, control over the virus population was lost due to multiplication of drug-resistant strains. A result of genetic modifications is the formation of virus mutations in transformation conditions $\gamma_{lb}$ resulting from their properties which belong to set $Pr$. In volumes $V_{\gamma ji(\gamma < \gamma_{lb})}$ formed by the drug, the concentration of transformation conditions $\gamma$ is smaller than that required for the transformation of so mutated viruses. Similar effects are observed in presently used, much more efficient highly active antiretroviral therapy (HAART). It includes the interactions of protease inhibitors combined with reverse transcriptase inhibitors. A mathematical model of the combined therapy was developed by Perelson et al. [29]. The same relations can be generated from Eq. (81). Objects from set $N$ according to their properties may be vulnerable to any of the means used in combination therapy (usually five) or to none. In the case of AIDS (acquired immunodeficiency syndrome), HAART does not result in patient's recovery despite a decrease of virus concentration below the detectability threshold. One of the reasons can be hiding of the virus in the milieu of limited immune response, such as brain and testes. They are unreachable for therapy just like memory lymphocytes and dendritic cells which are also a target of HIV attack. In the phenomenological model (Fig. 3), regions inaccessible for transformations are volumes $V_\delta$ (brain, testes) and $V_{ac}$ (dendritic cells, memory lymphocytes). Objects can remain in them until producing mutation which is resistant to the applied drug combination, capable of regaining the whole space V.

Now, to model an artificial immune system, the shape space concept proposed by Perelson and Ostera [30] is often used. Antigen and antibody were determined as a point in the L-dimensional space of complementary traits. The notion of threshold $\varepsilon$ determines the level of imperfectness of fitting of the antigen-antibody activation.

The idea of the transformation process preceded by a process of forming objects from a set $N$ (Section 2.2) can be used to model the epidemic taking into account such factors as vaccinations or immunization. The solution achieve for the simple SIR model (division of the individuals: susceptible, infecting and convalescents with acquired resistance) is consistent with a simple epidemic model developed by Kermack and McKendrick [31].


# 6. Summary

The presented theory relates to the physical, chemical and biological processes of random transformation of dispersed matter. It has the interdisciplinary significance allowing the phenomenological and mathematical modeling of mass transfer processes in many areas.

These include among others: industrial technology, ecology and environment protection, medicine, veterinary medicine, immunology, oncology, epidemiology, hygiene and agriculture. Specific descriptions of the processes create the possibility of linking phenomena, mechanisms and factors determining the process. Obtaining a correct modeling effect must be preceded by knowledge and deep understanding of the nature of the problem. However, full success depends on the proper, conducted on the basis of areas considered process, final interpretation of the results. The presented general concept of transformation of matter can be used both to study and describe the processes as well as for their management and control. It systematizes a range of knowledge concerning the transformation of dispersed matter whose nature has not previously been combined into unity. Together with the given methodology of building phenomenological and mathematical models it makes a platform on which it seems possible to achieve significant scientific development among others through analogies, critical comparisons and transfer of knowledge.

## Acknowledgements

The study was carried out within the frames of the grant W-10/1/2012/Dz. St.

## Author details

Marek Solecki*

Address all correspondence to: solecki@wipos.p.lodz.pl, msolecki@toya.net.pl

Department of Process Equipment, Faculty of Process and Environmental Engineering, Lodz University of Technology, Poland

## References

[1] Volterra V. Variazioni e fluttuazioni del numero d'individui in specie animal conviventi. Mem Acad Lincei Roma (1926) 2, 31-113; Volterra V. Variation and fluctuations of the number of individuals in animal species living together, in R.N. Chapman (Ed.), Animal Ecology, McGraw-Hill, New York (1931) 409-448.

[2] Lotka A.J. Undamped oscillations derived from the law of mass action. Journal of the American Chemical Society (1920) 42, 1595-1599.

[3] Chick H. An investigation of the laws of disinfection. Journal of Hygiene (1908) 8, 92-158.

[4] Watson H.E. A note on the variation of rate of disinfection with the change in the concentration of disinfectant. Journal of Hygiene (1908) 8, 536-542.

[5] Michaelis L., Menten M.I. Die Kinetik der Invertinwirkung. Biochemische Zeitschrift (1913) 49, 333-369.

[6] Hetherington P.J., Follows M., Dunnill P., Lilly M.D. Release of protein from baker's yeast (Saccharomyces cerevisiae) by disruption in an industrial homogeniser. Transactions of the Institution of Chemical Engineers (1971) 49, 142-148.

[7] Follows M., Hetherington P.J., Dunnill P., Lilly M.D. Release of enzymes from bakers' yeast by disruption in an industrial homogenizer. Biotechnology and Bioengineering (1971) 13, 549-560.

[8] Melendres A.V., Honda H., Shiragami N., Unno H. Enzyme release kinetics in a cell disruption chamber of a bead mill. Journal of Chemical Engineering of Japan (1993) 26(2), 148-152.

[9] Ho D.D., Neumann A.U., Perelson A.S., Chen W., Leonard J.M., Markowitz M., Rapid turnover of plasma virions and CD4 lymphocytes in HIV-1 infection. Nature (1995) 373, 123-126.

[10] Solecki, M. Mechaniczna dezintegracja komórek mikroorganizmów. Zeszyty naukowe Politechniki Łódzkiej (2012) 1114(421), 1-95.

[11] histi, Y, & Moo-Young, M. Disruption of microbial cells for intracellular products. Enzyme and Microbial Technology (1986) 8, 194-204.

[12] Heim A., Kamionowska U., Solecki M., The effect of microorganism concentration on yeast cell disruption in a bead mill. Journal of Food Engineering (2007) 83, 121-128.

[13] Heim A., Solecki M., Disintegration of microorganisms in a circulating bed of balls. Proceedings of the 3rd World Congress on Particle Technology, CD-ROM, ISBN 0-85295-401-8Brighton UK, 6-9 July (1998) 1-10.

[14] Solecki M. The release of compounds from microbial cells, in Nakajima H. (Ed.), Mass Transfer - Advanced Aspects, ISBN 978-953-307-636-2, InTech, Rijeka, 26 (2011) 595-618. Available from: http://www.intechopen.com/articles/show/title/the-release-of-compounds- from-microbial-cells, 26(2011) 595-618.

[15] Sauer T., Robinson C. W. and Glick B. R. Disruption of native and recombinant Escherichia coli in a high-pressure homogenizer. Biotechnology and Bioengineering (1989) 33, 1330-1342.

[16] Keshavarz Moore, E, Hoare, M, & Dunnill, P. Disruption of baker's yeast in a high-pressure homogenizer: new evidence on mechanism. Enzyme and Microbial Technology (1990) 12, 764-770.

[17] Engler, C. R. Cell disruption by homogenizer. In: Separation Processes in Biotechnology (Asenjo J.A., Ed.), Marcel Dekker, New York (1990) 95-105.

[18] Lander R., Manger W., Scouloudis M., Ku A., Davis C., Lee A. Gaulin homogenization; a mechanistic study. Biotechnology Progress (2000) 16, 80-85.

[19] Milner H.W., Lavrence N. S. and French G. S. Colloidal dispersion of chloroplast material. Science (1950) 111, 633-634.

[20] Wimpenny W.T. Breakage of micro-organizm, Process Biochemistry (1967) July 2(7), 41-44.

[21] Emanuel C.F. and Chaikoff J.L. An hydraulic homogenizer for the controlled release of cellular components from various tissues. Biochimica biophysica Acta (1957) May, 24, 254-261.

[22] Daulah M.S. Mechanism o disintegration of biological cells in ultrasonic cavitation. Biotechnology and Bioengineering (1977) 19, 649-660.

[23] James C.J., Coakley W.T., Hughes D.E., Kinetics of protein release from yeast sonicated in batch and flow systems at. 20 kHz. Biotechnology and Bioengineering (1972) 14, 33-42.

[24] Esche R., Untersuchung der Schwingungskavitation in Flüssigkeiten. Acoustica(Akustische Beihefte), (1952) 2, 208-219.

[25] Wang D.I.C., Cooney C.L., Demain A.L., Dunnill P., Humphrey A.E., Lilly M.D., Fermentation and Enzyme Technology, in Wiley J. (Ed.), New York, (1979).

[26] олмогоров, А. Н. Локальная структура турбулентности в несжимаемой вязкой жидкости при очень больших числах Рейнольдса. Доклады АН СССР (1941) 30 299-303.

[27] Колмогоров, А. Н. Рассеяние энергии при локально изотропной турбулентности. Доклады АН СССР (1941) 32, 19-21.

[28] Hunt N.K., Marianas B.J. Inactivation of Escherichia coli with ozone: chemical and inactivation kinetics. Water Research (1999) 33(11), 2633-2641.

[29] Perelson A.S., Neumann A.U., Markowicz M., Leonard J.M., Ho D.D. HIV-1 dynamics in vivo: Virionclearance rate, infected life-span, and viral generation time. Science (1996) 271, 1582-1586.

[30] Perelson A.S., Ostera G.F. Theoretical studies of clonal selection: Minimal antibody repertoire size and reliability of self-non-self discrimination. Journal of Theoretical Biology (1979) 81, 645-670.

[31] Kermack W.O., McKendrick A.G. Contributions to the mathematical theory of epidemics. Proceedings of the Royal Society Lond. A (1927) 115, 700-721.

# Successive Linearization of Heat and Mass Transfer over an Unsteady Stretching Permeable Surface in the Presence of Thermal Radiation and a Variable Chemical Reaction

Stanford Shateyi and Sandile S. Motsa

Additional information is available at the end of the chapter

## 1. Introduction

Theoretical studies of viscous incompressible flows over continuous stretching surfaces through a quiescent fluid have their origins in the pioneering work of Crane 1970. These types of flows occur in many industrial processes, such as in glass fibre production, food stuff processing reactor fluidization, and transpiration cooling. The prime aim in almost every extrusion is to maintain the surface quality of the extrudate. The pioneering works of Crane have been extended by many researchers to explore various aspects of the flow and heat and mass transfer occurring in infinite domains of the fluid surrounding the stretching sheet, [Liu and Andersson 2008; Abd EL-Aziz 2009; Abel and Mahesha 2008; Shateyi and Motsa 2009; Ziabakhsh et al. 2010; Motsa and Sibanda 2011], among others.

Many practical diffusive processes involve molecular diffusion of species in the presence of chemical reaction within and/or at the boundary. Chemical reaction can tremendously alter diffusion rates in convective heat and mass transfer processes. The effect of a chemical reaction depends on whether the reaction is heterogenous or homogeneous, as well as whether it occurs at an interface or a single phase volume reaction. We call a reaction of order $n$, if the reaction rate is proportional to the $n$th power of the concentration. The study of chemical reaction processes is useful for improving a number of chemical technologies such as polymer production and food processing. Various aspects of this problem have been studied by some researchers (Alam et al. 2009; Shateyi et al. 2010; Cortel 2007; Alam and Ahammad 2011; Afify and Elgazery 2012).

There has been much interest in the study of radiative heat transfer flows due to the effect of radiation on performance of many engineering systems applying electrically conducting fluids. Many engineering processes such as nuclear plants, gas turbines, satellites and space vehicles, take place at high temperatures and thus the effect of thermal radiation cannot be ignored. Recently, flow, heat and/or mass transfer with thermal radiation have been studied by (Abd El-Aziz 2008; Shateyi and Motsa 2009; Pal and Mondal 2011).

In this chapter, we explore the semi analytic solution of the non linear heat and mass transfer over an unsteady stretching permeable surface with prescribed wall conditions in the presence of thermal radiation and a non-uniform chemical reaction. The proposed method of solution employed in this work is based on an extension of the quasilinearization method (QLM) that was initially proposed in Bellman and Kalaba (1965). This method employs Taylor series linearization to convert a nonlinear two-point boundary value problem into an iterative scheme of solution which can be integrated using various numerical techniques. Mandelzweig and his co-workers (see for example, Krivec et al. 1991; Mandelzweig 2005; Krivec and Mandelzweig 2008, among others) have recently extended the application of the QLM to a wide variety of nonlinear BVPs and established that the method converges quadratically. The integration of the QLM iteration scheme is performed using the Chebyshev spectral collocation method Canuto et al. 1988. Several studies (see for example, Awad et al.2011; Makukula et al. 2010a; Makukula et al. 2010b; Makukula et al. 2010c; Makukula et al. 2010d; Motsa 2011; Motsa and Shateyi 2010; Motsa 2011 and Shateyi Motsa 2011) have shown that blending the Chebyshev spectral method with iteration schemes like the QLM results in a highly accurate method which can be used to solve a wide variety of nonlinear boundary value problems.

In this work we present new iteration schemes which are based on systematically extending the QLM. The objective of this work is to demonstrate that the convergence rate of the QLM can be significantly improved by using the proposed iterations schemes.

## 2. Mathematical formulation

The study investigates the unsteady laminar boundary layer in a quiescent viscous incompressible fluid on a horizontal sheet which comes through a slot at the origin. At $t = 0$, the sheet is stretched with velocity $U_w(x, t)$ along the x-axis, keeping the origin in the fluid of ambient temperature $T_\infty$ and concentration $C_\infty$. The RosseLand approximation is used to describe the radiative heat flux in the energy equation. We also assume a variable chemical reaction.

The velocity, temperature and concentration fields in the boundary layer are governed by the two dimensional boundary layer equations for mass, and chemical species given by

$$\frac{\partial u}{\partial x} + \frac{\partial v}{\partial y} = 0, \tag{1}$$

$$\frac{\partial u}{\partial t} + u\frac{\partial u}{\partial x} + v\frac{\partial u}{\partial y} = \nu\frac{\partial^2 u}{\partial y^2}, \tag{2}$$

$$\frac{\partial T}{\partial t} + u\frac{\partial T}{\partial x} + v\frac{\partial T}{\partial y} = \alpha_0\frac{\partial^2 T}{\partial y^2} - \frac{1}{\rho c p}\frac{\partial q_r}{\partial y}, \tag{3}$$

$$\frac{\partial C}{\partial t} + u\frac{\partial C}{\partial x} + v\frac{\partial C}{\partial y} = D_m\frac{\partial^2 C}{\partial y^2} - K_l(C - C_\infty)^n, \tag{4}$$

Where $u, v$ are the velocity components in the $x$ and $y$ directions,respectively, $\nu$ is the kinematic viscosity,$g$ is the acceleration due to gravity, $\rho$ is the density of the fluid, $T$ and $T_\infty$

are the temperature of the fluid inside the thermal boundary layer and of the fluid in the free stream, respectively while $C$ and $C_\infty$ are the corresponding concentrations, $\alpha_0$ is the thermal diffusivity, $c_p$ is the specific heat at constant pressure, $D_m$ is the mass diffusivity and $q_r$ is the radiative heat flux.

The boundary conditions are given as follows:

$$u = U_w, \quad v = V_w, \quad T = T_w, \quad C = C_w \quad \text{at} \ y = 0, \tag{5}$$

$$u \to 0, \quad T \to T_\infty, \quad C = C_\infty, \quad \text{as} \ y \to \infty. \tag{6}$$

The stretching velocity $U_w(x,t)$, the surface temperature $T_w(x,t)$ and the surface concentration are assumed to be of the form: $U_w(x,t) = a/(1-ct)$, $T_w(x,t) = T_\infty + bx/(1-ct)$, $C_w(x,t) = C_\infty + bx/(1-ct)$, where a, b and c are positive constants with (ct<1), and both a and c have dimension reciprocal time.

The radiative heat flux $q_r$ is described by the Rosseland approximation such that

$$q_r = -\frac{4\sigma^*}{3K} \frac{\partial T^4}{\partial y}, \tag{7}$$

where $\sigma^*$ and $K$ are the Stefan-Boltzman constant and the mean absorption coefficient, respectively. Following Chamkha (1997), we assume that the temperature differences within the flow are sufficiently small so that the $T^4$ can be expressed as a linear function after using Taylor series to expand $T^4$ about the free stream temperature $T_\infty$ and neglecting higher order terms. This results in the following approximation

$$T^4 \approx 4T_\infty^3 T - 3T_\infty^4 \tag{8}$$

Using equations (7) and (8) in equation (3) we obtain

$$\frac{\partial q_r}{\partial y} = -\frac{16\sigma^* T_\infty^3}{3K} \frac{\partial^2 T}{\partial y^2}.$$

## 2.1. Similarity solutions

Now we introduce the following dimensionless functions of $f, \theta$ and $\phi$ and similarity variable $\eta$ (Ishak et al., 2009).

$$\eta = \left(\frac{U_w}{vx}\right)^{\frac{1}{2}} y, \ \psi = (vxU_w)^{\frac{1}{2}} f(n), \ \theta(n) = \frac{T - T_\infty}{T_w - T_\infty}, \ \phi = \frac{C - C_\infty}{C_w - C_\infty}, \tag{9}$$

where

$\psi(x,y,t)$ is a stream function defined as $u = \frac{\partial \psi}{\partial y}$ and $v = -\frac{\partial \psi}{\partial x}$

The governing equations are then transformed into a set of ordinary equations and associated boundary conditions as given below:

$$f''' + ff'' - (f')^2 - A(f' + \frac{\eta}{2}f'') = 0, \tag{10}$$

$$(3R + 4)\theta'' + 3RPr[f\theta' - 2f'\theta - \frac{A}{2}(3\theta + \eta\theta')] = 0, \tag{11}$$

$$\phi'' + Sc[f\phi' - 2f'\phi - \frac{A}{2}(3\phi + \eta\phi')] - ScK\phi^n = 0, \tag{12}$$

where A=c/a is the component that measures the unsteadiness, $Pr = \nu/\alpha$ is the Prandtl number, $R = 16\sigma T_\infty^3/3Kk$ is the radiation parameter, $Sc = \nu/D_m$ is the Schmidt number and $K = K_l(C_w - C_\infty)^{n-1}x/U_w(x,t)$ is the local chemical reaction parameter.

The boundary conditions are:

$$f(0) = f_w, \; f'(0) = 1, \; \phi(0) = 1, \; \theta(0) = 1, \tag{13}$$

$$f'(\infty) = 0, \; \theta(\infty) = 0, \; \phi(\infty) = 0, \tag{14}$$

with $f_w < 0$ and $f_w > 0$ corresponding to injection and suction, respectively.

## 3. Method of solution

To solve the governing system of equations (12 - 14) we observe that equation (10) depends on $f(\eta)$ only. Thus, it can be solved independently of the other equations in the system. The solution for $f(\eta)$ is then substituted in equations (11) and (12) which can also be solved for $\theta$ and $\phi$ separately. We begin by obtaining the solution for $f(\eta)$. We assume that an estimate of the the solution of (10) is $f_\gamma$. For convenience, we introduce the following notation

$$f_0 = f, \; f_1 = f', \; f_2 = f'', \; f_3 = f'''. \tag{15}$$

In terms of the new variables (15), equation (10) can be written as

$$L[f_0, f_1, f_2, f_3] + N[f_0, f_1, f_2, f_3] = 0, \tag{16}$$

where

$$L[f_0, f_1, f_2, f_3] = f_3 - A(f_1 + \frac{\eta}{2}f_2), \quad N[f_0, f_1, f_2, f_3] = f_0 f_2 - f_1^2 \tag{17}$$

We introduce the following coupled system,

$$L[f_0, \ldots, f_3] + N(f_{0,\gamma}, \ldots, f_{3,\gamma}) + \sum_{s=0}^{3}(f_s - f_{s,\gamma})\frac{\partial N}{\partial f_s}(f_{0,\gamma}, \ldots, f_{3,\gamma}) + G(f_0, \ldots, f_3) = 0, \tag{18}$$

$$G(f_0, \ldots, f_3) = N(f_0, \ldots, f_3) - N(f_{0,\gamma}, \ldots, f_{3,\gamma}) - \sum_{s=0}^{3}(f_s - f_{s,\gamma})\frac{\partial N}{\partial f_s}(f_{0,\gamma}, \ldots, f_{3,\gamma}). \tag{19}$$

Successive Linearization of Heat and Mass Transfer over an Unsteady Stretching Permeable
Surface in the Presence of Thermal Radiation and a Variable Chemical Reaction

37

Note that when equations (18) and (19) are added, we obtain equation (16). Separating the
known and unknown variables, equation (18) can be written as

$$L[f_0,\dots,f_3] + \sum_{s=0}^{3} f_s \frac{\partial N}{\partial f_s}(f_{0,\gamma},\dots,f_{3,\gamma}) + G(f_0,\dots,f_3) = 0 \tag{20}$$

where

$$H(f_{0,\gamma},\dots,f_{3,\gamma}) = \sum_{s=0}^{3} f_{s,\gamma} \frac{\partial N}{\partial f_s}(f_{0,\gamma},\dots,f_{3,\gamma}) - N(f_{0,\gamma},\dots,f_{3,\gamma}) \tag{21}$$

We use the quasilinearization method (QLM) of Bellman and Kalaba (1965) to solve equation
(20). The QLM determines the $(i+1)$th iterative approximation $f_{s,i+1}$ as the solution of the
differential equation

$$L[f_{0,i+1},\dots,f_{3,i+1}] + \sum_{s=0}^{3} f_{s,i+1} \frac{\partial N}{\partial f_s}(f_{0,\gamma},\dots,f_{3,\gamma}) + G(f_{0,i},\dots,f_{3,i}) \tag{22}$$

$$+ \sum_{s=0}^{3} (f_{s,i+1} - f_{s,i}) \frac{\partial G}{\partial f_s}(f_{0,i},\dots,f_{3,i}) = H(f_{0,\gamma},\dots,f_{3,\gamma}).$$

Separating the unknowns $f_{s,i+1}$ from the known functions $f_{s,i}$ yields

$$L[f_{0,i+1},\dots,f_{3,i+1}] + \sum_{s=0}^{3} \left[\frac{\partial N}{\partial f_s}(f_{0,\gamma},\dots,f_{3,\gamma}) + \frac{\partial G}{\partial f_s}(f_{0,i},\dots,f_{3,i})\right] f_{s,i+1} = \tag{23}$$

$$\sum_{s=0}^{3} f_{s,i} \frac{\partial G}{\partial f_s}(f_{0,i},\dots,f_{3,i}) - G(f_{0,i},\dots,f_{3,i}) + H(f_{0,\gamma},\dots,f_{3,\gamma}),$$

subject to

$$f_{0,i+1}(0) = 0, \quad f_{1,i+1}(0) = 1, \quad f_{1,i+1}(\infty) = 0, \tag{24}$$

We assume that $f_{s,0}$ is obtained as a solution of the linear part of equation (20) given by

$$L[f_{1,0},\dots,f_{3,0}] + \sum_{s=0}^{3} f_{s,0} \frac{\partial N}{\partial f_s}(f_{0,\gamma},\dots,f_{3,\gamma}) = H(f_{0,\gamma},\dots,f_{3,\gamma}), \tag{25}$$

which yields the iteration scheme

$$L[f_{0,r+1},\dots,f_{3,r+1}] + \sum_{s=0}^{3} f_{s,r+1} \frac{\partial N}{\partial f_s}(f_{0,r},\dots,f_{3,r}) = H(f_{0,r},\dots,f_{3,r}). \tag{26}$$

It can easily be shown that equation (26) is the standard QLM iteration scheme for solving (16).

When $i = 0$ in (23) we can approximate $f_s$ as

$$f_s \approx f_{s,1}. \tag{27}$$

Thus, setting $i = 0$ in (23) we obtain

$$L[f_{0,1}, \ldots, f_{3,1}] + \sum_{s=0}^{3} \left[ \frac{\partial N}{\partial f_s}(f_{0,\gamma}, \ldots, f_{3,\gamma}) + \frac{\partial G}{\partial f_s}(f_{0,0}, \ldots, f_{3,0}) \right] f_{s,1} =$$

$$\sum_{s=0}^{3} f_{s,0} \frac{\partial G}{\partial f_s}(f_{0,0}, \ldots, f_{3,0}) - G(f_{0,0}, \ldots, f_{3,0}) + H(f_{0,\gamma}, \ldots, f_{3,\gamma}), \tag{28}$$

which yields the iteration scheme

$$L[f_{0,r+1}, \ldots, f_{3,r+1}] + \sum_{s=0}^{3} \left[ \frac{\partial N}{\partial f_s}(f_{0,r}, \ldots, f_{3,r}) + \frac{\partial G}{\partial f_s}(f_{0,r+1}^{(0)}, \ldots, f_{3,r+1}^{(0)}) \right] f_{s,r+1} =$$

$$\sum_{s=0}^{3} f_{s,r+1}^{(0)} \frac{\partial G}{\partial f_s}(f_{0,r+1}^{(0)}, \ldots, f_{3,r+1}^{(0)}) - G(f_{0,r+1}^{(0)}, \ldots, f_{3,r+1}^{(0)}) + H(f_{0,r}, \ldots, f_{3,r}) \tag{29}$$

where $f_{s,r+1}^{(0)}$ is the solution of

$$L[f_{0,r+1}^{(0)}, \ldots, f_{3,r+1}^{(0)}] + \sum_{s=0}^{3} f_{s,r+1}^{(0)} \frac{\partial N}{\partial f_s}(f_{0,r}, \ldots, f_{3,r}) = H(f_{0,r}, \ldots, f_{3,r}). \tag{30}$$

The general iteration scheme obtained by setting $i = m$ ($m \geq 2$) in equation (23), hereinafter referred to as scheme-$m$ is

$$L[f_{0,r+1}, \ldots, f_{3,r+1}] + \sum_{s=0}^{3} \left[ \frac{\partial N}{\partial f_s}(f_{0,r}, \ldots, f_{3,r}) + \frac{\partial G}{\partial f_s}(f_{0,r+1}^{(m-1)}, \ldots, f_{3,r+1}^{(m-1)}) \right] f_{s,r+1} =$$

$$\sum_{s=0}^{3} f_{s,r+1}^{(m-1)} \frac{\partial G}{\partial f_s}(f_{0,r+1}^{(m-1)}, \ldots, f_{3,r+1}^{(m-1)}) - G(f_{0,r+1}^{(m-1)}, \ldots, f_{3,r+1}^{(m-1)}) + H(f_{0,r}, \ldots, f_{3,r}) \tag{31}$$

where $f_{s,r+1}^{(m-1)}$ is obtained as the solution of

$$L[f_{0,r+1}^{(m-1)}, \ldots, f_{3,r+1}^{(m-1)}] + \sum_{s=0}^{3} \left[ \frac{\partial N}{\partial f_s} (f_{0,r}, \ldots, f_{3,r}) + \frac{\partial G}{\partial f_s} (f_{0,r+1}^{(m-2)}, \ldots, f_{3,r+1}^{(m-2)}) \right] f_{s,r+1}^{(m-1)} =$$

$$\sum_{s=0}^{3} f_{s,r+1}^{(m-2)} \frac{\partial G}{\partial f_s} (f_{0,r+1}^{(m-2)}, \ldots, f_{3,r+1}^{(m-2)}) - G(f_{0,r+1}^{(m-2)}, \ldots, f_{3,r+1}^{(m-2)}) + H(f_{0,r}, \ldots, f_{3,r}) \qquad (32)$$

The iterative schemes (26) and (31) can easily be solved using numerical methods such as finite differences, finite elements, Runge-Kutta based shooting methods or collocation methods. Several studies, (see for example, Awad et al.2011; Makukula et al. 2010a; Makukula et al. 2010b; Makukula et al. 2010c; Makukula et al. 2010d; Motsa 2011; Motsa and Shateyi 2010; Motsa 2011 and Shateyi Motsa 2011), have shown that the Chebyshev spectral collocation (CSC) method is very robust in solving iterations schemes of the type discussed in this work. The CSC method is based on approximating the unknown functions by the Chebyshev interpolating polynomials in such a way that they are collocated at the Gauss-Lobatto points defined as

$$z_j = \cos \frac{\pi j}{M}, \quad j = 0, 1, \ldots, M. \qquad (33)$$

where $M$ is the number of collocation points used (see for example Canuto et al. 1988, Trefethen 2000). For the convenience of numerical implementation, the domain $[0, \infty)$ is truncated as $[0, L_e]$ where $L_e$ is chosen to be a sufficiently large real number. In order to implement the method, the physical region $[0, L_e]$ is transformed into the region $[-1, 1]$ using the mapping

$$\eta = L_e \frac{z+1}{2}, \quad -1 \leq z \leq 1 \qquad (34)$$

The derivatives of $f$ at the collocation points are represented as

$$\frac{d^n f}{d\eta^n} = \sum_{k=0}^{M} \mathbf{D}_{kj}^2 f(z_k), \quad j = 0, 1, \ldots, M \qquad (35)$$

where $\mathbf{D} = 2D/L_e$, with $D$ being the Chebyshev spectral differentiation matrix (see for example, Canuto et al. 1988, Trefethen 2000). Thus, applying the CSC on the functions $f_s$ we obtain

$$f_s = \mathbf{D}^s \mathbf{F} \qquad (36)$$

where $\mathbf{F} = [f_0(z_0), f_0(z_1), \ldots, f_0(z_{M-1}), f_0(z_M)]^T$.

Thus, applying the spectral method, with derivative matrices on equation (26) and the corresponding boundary conditions gives the following matrix system

$$\mathbf{C}_r \mathbf{F}_{r+1} = \mathbf{H}_r \qquad (37)$$

with the boundary conditions

$$f_{0,r+1}(z_M) = 0, \quad \sum_{k=0}^{M} \mathbf{D}_{Mk} f_{0,r+1}(z_k) = 0, \quad \sum_{k=0}^{M} \mathbf{D}_{0k} f_{0,r+1}(z_k) = 1, \tag{38}$$

where

$$\mathbf{C}_r = \mathbf{D}^3 + \left(\mathbf{a}_{2,r} - \frac{A}{2}\eta_d\right)\mathbf{D}^2 + (\mathbf{a}_{1,r} - A)\mathbf{D} + \mathbf{a}_{0,r}. \tag{39}$$

The vector $\mathbf{H}_r$ corresponds to the function $H$ when evaluated at the collocation points and $\mathbf{a}_{s,r}$ $(s = 0, 1, 2)$ is a diagonal matrix corresponding to the vector of $a_{i,r}$ which is defined as

$$a_{s,r} = \frac{\partial N}{\partial f_s} \tag{40}$$

and $\eta_d$ is an $(M+1) \times (M+1)$ diagonal matrix of $\eta$. The boundary conditions (38) are imposed on the first, $M$th and $(M+1)$th rows of $C_r$ and $\mathbf{H}_r$ to obtain

$$\begin{pmatrix} \mathbf{D}_{0,0} & \mathbf{D}_{0,1} & \cdots & \mathbf{D}_{0,M-1} & \mathbf{D}_{0,M} \\ & & \mathbf{C}_r & & \\ \mathbf{D}_{M,0} & \mathbf{D}_{M,1} & \cdots & \mathbf{D}_{M,M-1} & \mathbf{D}_{M,M} \\ 0 & 0 & \cdots & 0 & 1 \end{pmatrix} \begin{pmatrix} f_{0,r+1}(z_0) \\ f_{0,r+1}(z_1) \\ \vdots \\ f_{0,r+1}(z_{M-2}) \\ f_{0,r+1}(z_{M-1}) \\ f_{0,r+1}(z_M) \end{pmatrix} = \begin{pmatrix} 1 \\ H_r(z_1) \\ \vdots \\ H_r(z_{N-2}) \\ 0 \\ 0 \end{pmatrix} \tag{41}$$

Starting from a suitable initial guess $f_{0,0}(\eta)$, the iteration scheme (41) can be used to iteratively give approximate solutions of the governing equation (10). The application of the CSC on the general iteration schemes (31) and (32) can be done in a similar manner for any value of $m$.

## 4. Results and discussion

In this section we present the results for the governing physical parameters of interest. In applying the Chebyshev spectral method described in the previous section $M = 100$ collocation points were used. The value of $L_e$ for numerically approximating infinity was chosen to be $Le = 20$. In order to assess the accuracy of the proposed iteration methods, the present numerical results were compared against results generated using the MATLAB routine bvp4c. For illustration purposes, results are presented for the first three iterations schemes obtained by setting $m = 0, 1, 2$.

In Table 1 we give a comparison between the results of scheme-0 against results generated using bvp4c. We observe that the QLM results converge very rapidly to the bvp4c results. It takes only three or four iterations to achieve an exact match that is accurate to order $10^{-8}$ for the selected parameters of $A$. We also observe in this table that stretching increases the absolute values of the skin friction.

| iter. | $A = 0$ | $A = 0.5$ | $A = 1$ | $A = 1.5$ | $A = 2$ |
|---|---|---|---|---|---|
| 1 | -1.000000000 | -1.166255146 | -1.317872855 | -1.455413216 | -1.581765234 |
| 2 | -1.000000000 | -1.167211134 | -1.320520326 | -1.459662889 | -1.587362322 |
| 3 | -1.000000000 | -1.167211513 | -1.320522065 | -1.459665895 | -1.587366111 |
| 4 | -1.000000000 | -1.167211515 | -1.320522065 | -1.459665895 | -1.587366111 |
| 5 | -1.000000000 | -1.167211517 | -1.320522065 | -1.459665895 | -1.587366111 |
| bvp4c | -1.000000000 | -1.167211517 | -1.320522065 | -1.459665895 | -1.587366111 |

**Table 1.** Comparison of the bvp4c values of $f''(0)$ at different values of $A$ for Scheme-0 (QLM)

| iter. | $A = 1$ | $A = 2$ | $A = 3$ | $A = 4$ |
|---|---|---|---|---|
| | | | Scheme-0 | |
| 1 | -1.215913833273934 | -1.508012270141307 | -1.750647835613831 | -1.963118416905635 |
| 2 | -1.315711765474042 | -1.586229425552791 | -1.816415412229436 | -2.022058898487201 |
| 3 | -1.320495400678920 | -1.587365641569519 | -1.816850675892770 | -2.020976536521761 |
| 4 | -1.320522063086358 | -1.587366111631070 | -1.816849325533777 | -2.020950025633773 |
| 5 | -1.320522064602713 | -1.587366111619306 | -1.816849325468859 | -2.020950025517386 |
| 6 | -1.320522064602713 | -1.587366111619306 | -1.816849325468859 | -2.020950025517386 |
| | | | Scheme-1 | |
| 1 | -1.315711765474042 | -1.586229425552791 | -1.816415412229436 | -2.022058898487201 |
| 2 | -1.320522063086358 | -1.587366111631070 | -1.816849325533777 | -2.020950025633773 |
| 3 | -1.320522064602713 | -1.587366111619306 | -1.816849325468859 | -2.020950025517386 |
| 4 | -1.320522064602713 | -1.587366111619306 | -1.816849325468859 | -2.020950025517386 |
| 5 | -1.320522064602713 | -1.587366111619306 | -1.816849325468859 | -2.020950025517386 |
| 6 | -1.320522064602713 | -1.587366111619306 | -1.816849325468859 | -2.020950025517386 |
| | | | Scheme-2 | |
| 1 | -1.315240928788073 | -1.585906261457893 | -1.816236653612290 | -2.021499891062478 |
| 2 | -1.320522059975388 | -1.587366111619187 | -1.816849325470411 | -2.020950025519239 |
| 3 | -1.320522064602713 | -1.587366111619306 | -1.816849325468859 | -2.020950025517386 |
| 4 | -1.320522064602713 | -1.587366111619306 | -1.816849325468859 | -2.020950025517386 |
| 5 | -1.320522064602713 | -1.587366111619306 | -1.816849325468859 | -2.020950025517386 |
| 6 | -1.320522064602713 | -1.587366111619306 | -1.816849325468859 | -2.020950025517386 |

**Table 2.** $f''(0)$ at different values of $A$ for scheme-0,1,2

| iter. | $A = 1$ | $A = 2$ | $A = 3$ | $A = 4$ |
|---|---|---|---|---|
| | | Scheme-0 | | |
| 1 | 0.104608231328780 | 0.079353841477999 | 0.066201489855027 | 0.057831608611751 |
| 2 | 0.004810299128671 | 0.001136686066515 | 0.000433913239423 | 0.001108872969816 |
| 3 | 0.000026663923793 | 0.000000470049787 | 0.000001350423911 | 0.000026511004376 |
| 4 | 0.000000001516355 | 0.000000000011764 | 0.000000000064918 | 0.000000000116388 |
| 5 | 0.000000000000000 | 0.000000000000000 | 0.000000000000000 | 0.000000000000000 |
| 6 | 0.000000000000000 | 0.000000000000000 | 0.000000000000000 | 0.000000000000000 |
| | | Scheme-1 | | |
| 1 | 0.004810299128671 | 0.001136686066515 | 0.000433913239423 | 0.001108872969816 |
| 2 | 0.000000001516355 | 0.000000000011764 | 0.000000000064918 | 0.000000000116388 |
| 3 | 0.000000000000000 | 0.000000000000000 | 0.000000000000000 | 0.000000000000000 |
| 4 | 0.000000000000000 | 0.000000000000000 | 0.000000000000000 | 0.000000000000000 |
| 5 | 0.000000000000000 | 0.000000000000000 | 0.000000000000000 | 0.000000000000000 |
| 6 | 0.000000000000000 | 0.000000000000000 | 0.000000000000000 | 0.000000000000000 |
| | | Scheme-2 | | |
| 1 | 0.005281135814640 | 0.001459850161413 | 0.000612671856568 | 0.000549865545092 |
| 2 | 0.000000004627325 | 0.000000000000119 | 0.000000000001552 | 0.000000000001853 |
| 3 | 0.000000000000000 | 0.000000000000000 | 0.000000000000000 | 0.000000000000000 |
| 4 | 0.000000000000000 | 0.000000000000000 | 0.000000000000000 | 0.000000000000000 |
| 5 | 0.000000000000000 | 0.000000000000000 | 0.000000000000000 | 0.000000000000000 |
| 6 | 0.000000000000000 | 0.000000000000000 | 0.000000000000000 | 0.000000000000000 |

**Table 3.** $f''(0)$ at different values of $A$

Tables 2 and 3 give the results of the comparison of the values of $f''(0)$ between three levels of the iteration schemes and their corresponding errors. In computing the errors it was assumed that the result corresponding to the 6th iteration is the converged solution. From numerical experimentation it was found that all the iteration schemes would have completely converged to a fixed value by the time the 5th or 6th iteration is used. The results from Table 2 and 3 clearly indicate that the the convergence to the solution progressively improves when you use the higher level iteration schemes. For instance, from Table 3 we note that it takes only 2 iterations to achieve full convergence in scheme-1 and scheme-2 compare to four iterations in scheme-0. This results demonstrates the improvement offered by the proposed new iteration scheme on the original quasilinearization method which corresponds to scheme-0.

In Figs. 1 - 7 we give illustrations showing the effect of the governing parameters on the flow properties. Unless otherwise specified, the sample illustrations were generated using $R = 1$, $Pr = 0.7$, $Sc = 1$, $K = 1$, $A = 1$. In Figs. 1 - 3 we show the velocity, temperature and concentration profiles for different values of $A$. In Fig. 1 we observe that the velocity $f'(\eta)$ is a monotonically decreasing function of the stretching parameter $A$. From Figs. 2 and 3, we observe that both the temperature and concentration distributions are reduced as values of the stretching parameter increase. The velocity, thermal and solutal boundary layer thicknesses all decrease as the values of $A$ increase. As a consequence the transition from

Successive Linearization of Heat and Mass Transfer over an Unsteady Stretching Permeable
Surface in the Presence of Thermal Radiation and a Variable Chemical Reaction

43

laminar flow to turbulent flow is delayed. This shows that stretching of surfaces can be used as a flow stabilizing mechanism.

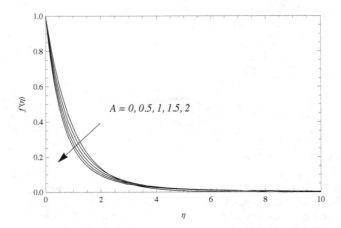

**Figure 1.** The variation of $A$ on the flow velocity, $f'(\eta)$

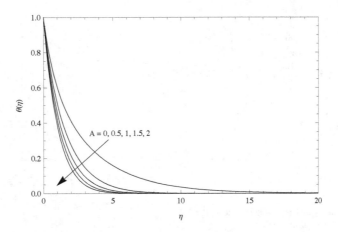

**Figure 2.** Effect of $A$ on the fluid temperature, $\theta(\eta)$

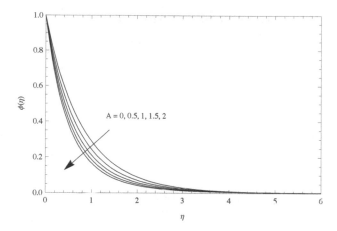

**Figure 3.** Effect of $A$ on the fluid concentration, $\phi(\eta)$

Fig. 4 depicts the effects of the Prandtl number on the temperature distributions. We observe that as $Pr$ increases, the temperature profiles and the thermal boundary layer thickness become smaller. This is because when $Pr$ increases, the thermal diffusivity decreases, leading to the decrease of the energy transfer ability that decreases the thermal boundary layer. The effect of thermal radiation $R$ on the temperature field is shown in Fig. 5. From this figure we see that the effect of increasing the thermal radiation parameter $R$ is to reduce the temperature profiles.

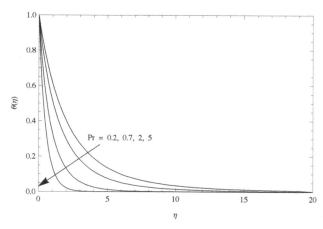

**Figure 4.** Temperature profiles for various values of $Pr$

Fig. 6 shows the dimensionless concentration profiles for different values of the Schmidt number $Sc$. We clearly see from this figure that the concentration boundary layer thickness

Successive Linearization of Heat and Mass Transfer over an Unsteady Stretching Permeable
Surface in the Presence of Thermal Radiation and a Variable Chemical Reaction

45

decreases as the Schmidt number $Sc$ increases. This phenomenon occurs because when $Sc$, the mass diffusivity decreases and the fluid becomes heavier. The effects of chemical reaction $K$ on the concentration distributions is displayed in Fig. 7. It should be noted here that physically positive values of $K$ implies the destructive reaction. We observe in this figure that an increase in the chemical reaction leads to the decrease in the concentration can profiles. This shows that diffusion rate can be tremendously altered by chemical reaction.

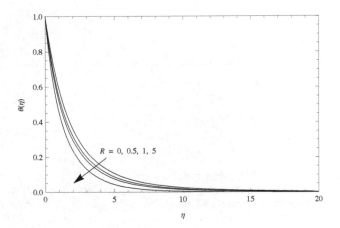

**Figure 5.** Effect of $R$ on the fluid temperature, $\theta(\eta)$

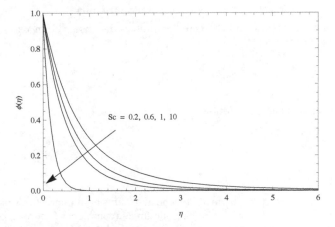

**Figure 6.** Effect of $Sc$ on the solute concentration, $\phi(\eta)$

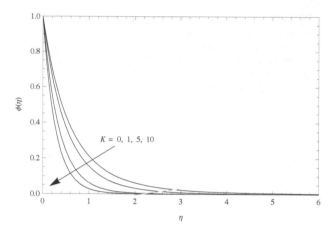

**Figure 7.** The solute concentration profiles for various values of $K$.

## 5. Conclusion

In this chapter we explored the semi-analytic solution of the non linear heat and mass transfer flow over an unsteady stretching permeable surface with prescribed wall conditions in the presence of thermal radiation and non-uniform chemical reaction. From the present investigation we may conclude the following:

1. Blending the QLM scheme with Chebyshev spectral collocation method leads to more accurate and faster convergence scheme.

2. Radiation significantly affect the fluid flow properties.

3. The diffusion rate is significantly altered by chemical radiation.

4. The velocity, temperature and concentration profiles decrease with increasing values of the stretching parameter $A$.

## Author details

Stanford Shateyi[1] and Sandile S. Motsa[2]

[1]University of Venda, South Africa
[2]University of KwaZulu-Natal, South Africa

## 6. References

[1] Abd El-Aziz, M. (2008). Thermal-diffusion and diffusion-thermo effects on combined heat and mass transfer by hydromagnetic three-dimensional free convection over a permeable stretching surface with radiation. *Physics Letters A*. 372, 263-372.

[2] Abd El-Aziz, M. (2009). Radiation effect on the flow and heat transfer over an unsteady stretching sheet, *Int. Commun. Heat Mass Transf,* doi:10.1016/j.icheatmasstransfer.2009.01.016.

[3] Abel, M.S. & Mahesha, N. (2009). Heat transfer in MHD viscoelastic fluid over a stretching sheet with variable thermal conductivity, non-uniform heat source and radiation. *Appl. Math. Modell.* 32, 1965-1983.

[4] Alam, M.S., Rahman, M.M., & Sattar, M.A. (2009). Transient Magnetohydrodynamic Free Convective Heat and Mass Transfer Flow with Thermophoresis past a Radiative Inclined Permeable Plate in the Presence of Variable Chemical Reaction and Temperature Dependent Viscosity. *Nonlinear Analysis: Modelling and Control.* 14., 1., pp 3-20.

[5] Alam, M.S., & Ahammad, M.U.(2011). Effects of variable chemical reaction and variable electric conductivity on free convective heat and mass transfer flow along an inclined stretching sheet with variable heat and mass fluxes under the influence of Dufour and Soret effects. *Nonlinear Analysis: Modelling and Control.* 16.1., pp 1-16.

[6] Afify, A.A., & Elgazery, N.S. (2012). Lie group analysis for the effects of chemical reaction on MHD stagnation-point flow of heat and mass transfer towards a heated porous stretching sheet with suction or injection. *Nonlinear Analysis: Modelling and Control.* 17.1. pp 1-15.

[7] Awad, F.G. Sibanda, P. Motsa, S.S. & Makinde, O.D. (2011). Convection from an inverted cone in a porous medium with cross-diffusion effects, *Computers and Mathematics with Applications,* 61, pp 1431–1441.

[8] Bellman. R.E., & R.E. Kalaba, (1965), Quasilinearization and Nonlinear Boundary-Value Problems, *Elsevier, New York.*

[9] Canuto, C, Hussaini, M.L, Quarteroni, A. & Zang, T.A. (1988). Spectral Methods in Fluid Dynamics. *Springer-Verlag, Berlin.*

[10] Chamkha, A. J., (1997). Hydromagnetic natural convetion from an isothermal inclined surface adjacent to a thermally stratified porous medium, *Int. J. Eng. Sci.,* 37, 10.11, pp 975 - 986.

[11] Crane, L. J. (1970). Flow past a stretching plate, *Z. Angew. Math. Physc.* 12, pp 645-647.

[12] Cortell, R. (2007). MHD flow and mass transfer of an electrically conducting fluid of second grade in a porous medium over a stretching sheet with chemically reactive species. *Chemical Engineering and Processing.* 46, pp 721-728.

[13] Ishak, A., Nazar, R., and Pop, I. (2009). Heat Transfer over an Unsteady Stretching Permeable Surface with Prescribed Temperature, *Nonlinear Analysis: Real World Applications,* 10, pp 2909-2913.

[14] Krivec, R.; Haftel, M.I. & Mandelzweig, V.B. (1991). Precisse nonvariational calculation of excited states of helium with the correction-function hyperspherical-harmonic method. *Physical Review A*. Vol. 44. No. 11. pp 7159-7164. December 1991.

[15] Krivec, R. & Mandelzweig, V.B. (2008). Quasilinearization approach to computations to computations with singular potentials. *Computer Physics Communications*. Vol. 179. pp 865-867.

[16] Chung Liu I & Andersson, H.L. (2008).Heat transfer in a liquid film on an unsteady stretching sheet, *International Journal of Thermal Sciences*, 47, pp 766-772.

[17] Makukula, Z.G. Sibanda, P.& Motsa, S.S. (2010a). A novel numerical technique for two-dimensional laminar flow between two moving porous walls, *Mathematical Problems in Engineering* Article ID 528956, pp 1-15; doi:10.1155/2010/528956.

[18] Makukula, Z.G. Sibanda, P.& Motsa, S.S. (2010b) A note on the solution of the von KÃạrmÃạn equations using series and Chebyshev spectral methods. *Boundary Value Problems*, Vol. 2010, Article ID 471793, pp 1-17 ; doi:10.1155/2010/471793.

[19] Makukula, Z.G., Sibanda, P., & Motsa, (2010c). On new solutions for heat transfer in a visco-elastic fluid between parallel plates. *International Journal of Mathematical Models and Methods in Applied Sciences*. 4, (4), pp 221 - 230.

[20] Makukula, Z.G., Motsa, S.S., & P. Sibanda. (2010). On a new solution for the viscoelastic squeezing flow between two parallel plates, *Journal of Advanced Research in Applied Mathematics*. 2, pp 31 - 38.

[21] Mandelzweig, V.B. (2005). Quasilinearization Method: Nonperturbative Approach to Physical Problems. *Physics of Atomic Nuclei*. Vol. 68. No. 7. pp 1228-1257.

[22] Motsa, S.S.& Shateyi,S. (2010). A New Approach for the Solution of Three-Dimensional Magnetohydrodynamic Rotating Flow over a Shrinking Sheet, *Mathematical Problems in Engineering*, vol. 2010, Article ID 586340, pp 1-15. doi:10.1155/2010/586340

[23] Motsa, S.S., & Sibanda, P., (2011). On the solution of MHD flow over a nonlinear stretching sheet by an efficient semi-analytical technique, *Int.J Numer.fluids*, pp 1-13.

[24] Motsa, S.S. (2011). New algorithm for solving non-linear BVPs in heat transfer, *International Journal of Modeling, Simulation & Scientific Computing*, 2(3), pp 355–373.

[25] Motsa, S.S. & Shateyi,S. (2011). Successive Linearisation Solution of Free Convection Non-Darcy Flow with Heat and Mass Transfer, *Advanced Topics in Mass Transfer*, Mohamed El-Amin (Ed.), pp 425-438, InTech Open Access Publishers.

[26] Pal, D., & Mondal, H., (2011). Hydromagnetic non-Darcy flow and heat transfer iver a stretching sheet in the presence of thermal radiation and Ohmic dissipation. *Commun Nonlinear Sci Numer Simulat*. 15. 1197-1209.

[27] Shateyi, S. Motsa, S.S., (2009). Thermal radiation effects on heat and mass transfer over an unsteady stretching surface, *Mathematical Problems in Engineering*, Vol. 2009, doi:10.1155/2009/965603.

[28] Shateyi, S., Motsa, S.S., & Sibanda, P. (2010). Homotopy analysis of heat and mass transfer boundary layer flow through a non-porous channel with chemical reaction and heat generation. *The Canadian Journal of Chemical Engineering*. 88. pp 975-982.

[29] Shateyi, S., Motsa, S.S., (2010). Variable viscosity on magnetohydrodynamic fluid flow and heat transfer over an unsteady stretching surface with Hall effect, *Boundary Value Problems*, Vol. 2010, Article ID 257568, pp 1-20, doi:10.1155/2010/257568

[30] Trefethen, L.N., Spectral Methods in MATLAB, SIAM, 2000

[31] Zibakhsh, Z., Domairry, G., Mozaffari, M., & Mahbobifar, M. (2010). Analytical solution of heat transfer over an unsteady stretching permeable surface with prescribed wall temperature. *Journal of the Taiwan Institute of Chemical Engineers*. 41, pp 169-177.

# Chapter 3

# Continuous Chromatography Modelling with 2D and 3D Networks and Stochastic Methods – Effects of Porous Structure and Solute Population

Leôncio Diógenes T. Câmara, Jader Lugon  Junior, Flávio de Matos Silva, Guilherme Pereira de Oliveira, Lídice Camps Echevarria, Orestes Llanes Santiago and Antônio J. Silva Neto

Additional information is available at the end of the chapter

## 1. Introduction

The modelling of separation chromatographic processes reported in the literature is, in general, related to macroscopic approaches for the phenomenological representation of the mass transfer mechanisms involved. In such models the microscopic aspects of the porous medium structure, related to the separation mechanisms, are incorporated implicitly, limiting the quality of the representation of the separation systems, which are strongly influenced by the micro porous adsorbent.

The modelling of fluid flow in porous media through the application of interconnected networks, which considers the global result from a system of interconnected microscopic elements, is related to the concepts of percolation theory.

The classical macroscopic models of chromatography have limitations in representing the structural parameters of the solid adsorbents, such as topology and morphology, as well as population effects of the molecules in the liquid phase, i.e. multi-molecules movement. Such important microscopic properties can be represented applying interconnected network models which can lead to a better understanding of the phenomenological aspects that contribute to the separation mechanisms in micro-porous media.

In the simulation of the molecules flow through the column porous medium, a stochastic approach is utilized to represent the adsorption, diffusion and convection phenomena.

The molecules can move freely in the network structure, from one neighbor site to another, being respected the requirement that the final position is not occupied by another molecule. Therefore, two molecules cannot occupy the same network site.

This chapter is dedicated to the modelling of continuous chromatography with a network approach combined with Monte Carlo like random walk stochastic methods. The porous structure of the solid adsorbent phase of the chromatographic column is represented by two and three dimensional networks, respectively square and cubic lattices (Oliveira *et al.*, 2008; Biasse *et al.* 2010), in which population effects are taken into account, being related to the movement of multi-molecules modeled by stochastic phenomena of adsorption, desorption, diffusion and advection.

The use of network models to study chromatographic separation processes has been observed in the literature with different techniques and applications (Loh & Wang, 1995; Kier *et al.*, 2000; Loh & Geng, 2003; Geng & Loh, 2004; Bryntensson, 2002; Oliveira et al., 2008, Biasse et al., 2010). In the work of Kier *et al.* (2000), a square network model was applied in the representation of the chromatographic column utilizing a cellular automata approach. The authors assumed arbitrarily pre-defined probabilities for the particles motion and interactions among them. Loh & Geng (2003) applied a cubic network model of interconnected cylindrical pores in the study of chromatographic systems of perfusion. Topological and morphological aspects, such as connectivity and pore size distribution, were analyzed, being observed a great influence of such porous adsorbent media characteristics on the mass transfer of the phenomena studied. In Geng & Loh (2004), the porous structure of the column was modeled considering three different Gaussian distributions of pore size, in order to represent the macro-pores, the micro-pores and the interstitial pores.

In this chapter, the fluid advection is assumed to be the main factor contributing to the molecules movement in the network structure. Such assumption is reasonable since the fluid movement in the chromatographic column comes from the external driving force provided by the pumping system. Two pore dimensions are assumed in the porous structure of adsorbent, the small and large cavities, which leads to steric and non-steric effects, respectively, due to the required space for the molecules movement.

The application of interconnected network models combined with stochastic phenomena of adsorption, diffusion and advection represents the main dynamical behaviors of the chromatographic processes of separation. The multi-molecules population effects allow the study of the dynamics of percolation through the chromatographic column, being therefore possible to evaluate the influence of molecules concentration on the mass transport phenomena along the chromatographic column.

## 2. Phenomenological and structural modelling

In the adsorption phenomenon, the molecule arriving at a new site can be adsorbed according to a probability of adsorption ($p_{ads}$). In this case, the system generates a random number R, from a uniform distribution between 0 and 1, being it compared to a pre-defined adsorption probability $(p_{ads})$. The molecule is then adsorbed at the new site if the random number R is smaller than the adsorption probability $(R < p_{ads})$. The same procedure is applied in the desorption of a previously adsorbed molecule, with a desorption probability $(p_{des})$. The relation between the adsorption and desorption probabilities considered in the simulations is

$$p_{ads} + p_{des} = 1, \quad (0 \leq p_{ads} \leq 1) \tag{1}$$

The adsorption and desorption phenomena are represented by

$$A + s \underset{k_2}{\overset{k_1}{\underset{\leftarrow}{\rightarrow}}} A.s \tag{2}$$

in which a molecule of solute $A$ can be adsorbed at or desorbed from a site s according to a kinetic constant of adsorption $(k_1)$ or desorption $(k_2)$, respectively. From Eq. (2) it can be observed that the adsorption rate is proportional to the concentration of both solute A and vacant sites s.

The simulation of the dynamic process of chromatographic separation in adsorption columns was performed combining the network modelling of the column porous structure with the stochastic modelling of molecules movement and interactions percolating the system.

In the schematic representation of the adsorption column shown in Fig. 1, the symbols $C_0$ and C represent the solute concentrations at the entrance and at the exit of the column, respectively.

A square network was used to model surface phenomenon (studying equilibrium isotherms) and a cubic one to model the diffusion phenomenon. The porous medium of the chromatography column was represented by a two or three-dimensional cubic network model of interconnected sites (Vide Fig. 2). In such structure the connectivity, i.e. the number of neighbors connected to each site, is equal to four (2D) or six (3D). Each intersection node or site in the network corresponds to a potential adsorption location, in which the solute molecule may be adsorbed, being permitted only one adsorbed molecule per intersection.

In Fig. 2 one can see the graphical representation of both models used with the percolation threshold values $(p_{cs})$ considered over the site approach. The percolation threshold is the minimum probability of site occupation that represents the percolation through the whole system, becoming a cluster that goes from one extreme of the network to the other.

Continuous Chromatography Modelling with 2D and 3D Networks and Stochastic Methods – Effects of Porous Structure and Solute Population

53

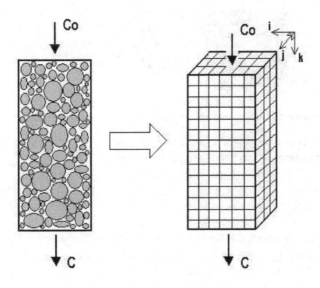

**Figure 1.** Chromatographic column represented by a three-dimensional cubic network.

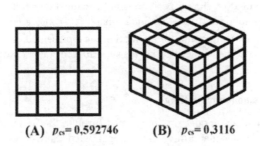

(A) $p_{cs}= 0,592746$    (B) $p_{cs}= 0,3116$

**Figure 2.** Graphical representation of two-dimension square network (A) and cubic network (B) with respective percolation threshold values ($p_{cs}$)

## 2.1. Adsorption isotherm model

In this stage, molecules adsorption in stirred tanks was modeled, so that the adsorption surface was represented as a two-dimensional network. In Fig. 3 the process is represented, the liquid phase molecule can be adsorbed at the adsorbent surface material, being the latter represented as the network at the tank bottom.

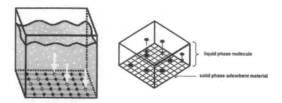

**Figure 3.** Stirred tank represented by a two-dimensional square network.

In this model it is assumed that the liquid phase molecules can be adsorbed at the network according to an adsorption probability ($p_{ads}$), defined as

$$p_{ads} = \frac{C_{i_t}}{C_{i_0}}$$

(3)

where $C_{i_t}$ and $C_{i_0}$ represent the liquid phase molecules concentration at adsorption times previously prescribed as $t$ and $t=0$ respectively. It is known that at time $t=0$, the concentration is $C_{i_t} = C_{i_0}$. The network used was a 100 x 100 sites, so that it takes to a number of 10.000 adsorbent sites that can be occupied, or not, by the molecules in the system. For each time step ($t$) a liquid phase molecule can adsorb to the surface according to a previously prescribed adsorption probability ($p_{ads}$). The adsorption takes place if the random number ($R$), taken by a uniform random number generator, is lower than ($p_{ads}$), that is, $R < p_{ads}$. Another condition is that the random chosen position is not occupied by another molecule. The adsorbed molecules are subtracted from the liquid phase. It is considered a final time equal to $10^6$, so that a infinite adsorption time, or equilibrium, is represented. It is not considered the interaction between adsorbed molecules, i.e. the superficial diffusion processes are not taken into account.

## 2.2. Diffusion phenomenon modelling

The diffusion phenomenon model was preformed through the "random walks" technique both for the two-dimensional and three-dimensional percolation at the square network and at the cubic network, respectively. Using this procedure one can use diffusion parameters for chromatography porous media and topological properties from network models, and obtain a fundamental universal correlation for these complex phenomena. Two different diffusion phenomena were studied at this stage: the first assumes that a certain molecule may follow any direction at the network; and the second that assumes that a molecule is only able to diffuse in the axial direction. The second case establishes diffusion relations for the axial dispersion for a chromatographic column, which is a important phenomenon that draws a lot of attention in separation chromatographic processes. In general, on the macroscopic models, only dispersion at the axial direction is taken into account, neglecting the transversal dispersion, that makes the numerical solutions to be simpler. In the present modelling, both axial and transversal dispersion are implicit on the stochastic "random walk" model, since the molecule is able to diffuse on both directions.

Equation 4 represents the diffusion mechanism, in which the molecule can diffuse in ($n$) directions with equal diffusion probability ($p_i$) for each direction. In general, for the square and cubic network, with axial dispersion, ($n$) corresponds respectively to 4 and 6, but when axial dispersion is considered ($n$) corresponds to 3 and 5, respectively, being neglected the backward diffusion along the axial direction. In all cases, the moving probability for each direction ($i$) is considered to be equal.

$$\sum_{i=1}^{n} p_i = 1 \tag{4}$$

At first, a probability of occupation for network elements ($p_{oc}$) in each direction $i$ is considered to be equal. Simulations for $p_{oc} < 1$ were also done, in such a way to represent the porosity ($\varepsilon$) of the porous adsorbent media. A porosity represents the empty space fraction in the chromatographic column, being proportional to the occupation probability ($p_{oc} \propto \varepsilon$). In the case that $p_{oc} = 1$, it is considered that the column is completely empty, and therefore ($\varepsilon = 1$). Networks with $p_{oc} < 1$ were built randomly, being the existence of a certain site conditioned to the generation of a random number $R$, and the satisfaction of $R < p_{oc}$ condition.

Simulations for different time ranges were made (10, 25, 50 and 75), being monitored the distance ($d_p$) by the "random walk" and also the area ($A_p$), the latter counted in site units. In the case of 3D, the latter has a relationship with the volume ($V_d$), also counted in site units. For each time-step, a different molecule direction was randomly chosen, in a such way that the destination site was forced to exist; A random number of events $N = 10^6$ was performed in order to obtain a distance ($d_p$) by the "random walk" with smaller dispersion.

In the case of smaller molecular diffusion ($D_m$), it was observed a certain proportion $D_m \propto d_p$ with the distance, obeying to the power law (Stauffer and Aharony, 1992, Biasse et al., 2010).

$$d_p \propto c.t^k \tag{5}$$

where $c$ and k are constants and $t$ represents time.

## 2.3. Movement rules and steric effect

In the simulation of the molecules movements, four rules (MR – movement rules) are considered to be representative of the diffusion and advection mechanisms. Such rules are schematically represented in Fig. 4.

From Fig. 4 it can be observed that the movement rules are determined by the directions considered in each situation. In the movement rules (I) and (II), the molecules can move in all directions of the structure while in the movement rules (III) and (IV) the molecules cannot move in the direction 6, which corresponds to the movement against the longitudinal main stream flow direction, i.e. the upstream movement is not allowed. The MR-(II) and (IV) are similar to MR-(I) and (III), respectively, with a higher chance of movement in the direction 5 (longitudinal or axial direction).

The MR-(I) is considered to be representative of the diffusion mechanism of solute molecules through the porous structure.

This assumption is coherent as there is not a driving mechanism forcing the flow in any particular direction of the network. In this case, according to the first law of Fick (Bird et al., 2002), the solute flow is determined by the concentration gradient of molecules, without significant effects of external forces. The diffusion mechanism of MR-(I) is governed by

$$p_i = \frac{1}{6}, \quad i = 1, 2, \ \dots, \ 6 \tag{6}$$

in which $P_i$ represents the probability of the solute molecule to move in the i-th direction of the network. According to Eq. (6), the chance of the molecules to follow any direction (six directions) in the network is the same.

The MR-(II), (III) and (IV) are considered to be representative of the advective mechanism as these configurations favor the flow in the axial downstream direction of the column.

The advective mechanism MR-(II) is governed by

$$p_i = \frac{1 - p_5}{5}, \quad i = 1, 2, \ \dots, \ 6 (i \neq i_5), \quad p_5 > \frac{1}{6} \tag{7}$$

in which $p_5$ represents the probability of the solute molecule to move in the axial downward direction of the column. In this case, the chance of the solute molecule to move in the axial downward direction is greater than the other directions, and the probabilities related to the movements in the remaining directions are the same.

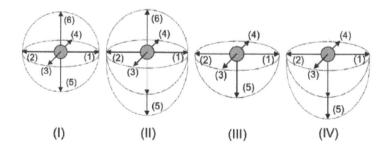

**Figure 4.** Movement rules (MR) for the solute molecules with the corresponding directions.

The advective mechanisms MR-(III) and (IV) are determined, respectively, by

$$p_i = \frac{1}{5}, \quad i = 1, 2, \ldots, 5 \tag{8}$$

$$p_i = \frac{1 - p_5}{4}, \quad i = 1, 2, \ldots, 4, \quad p_5 > \frac{1}{5} \tag{9}$$

From Eq. (8) it can be observed that the solute movement in the direction 6 is prohibited, having the same probability for the moves in the other directions. In Eq. (9) as in Eq. (7), the probability of moving in the direction 5 is greater than in the other directions.

The simulation of the flow of solute molecules was carried out according to the procedure described next. The molecules were introduced randomly at the column entrance (at the nodes with k = 1), maintaining the concentration constant at this section ($C_0$). The molecules concentration (C) at the column exit (k = 30) was calculated as the ratio of the number of molecules occupying the network intersections (sites) and the total number of intersections. In each step of the simulation, corresponding to a discrete time ($t^*$), all the network sites occupied by molecules were checked, being assumed for each molecule a direction of movement according to the MR to be followed. A molecule was kept at its original location if the new site, it was supposed to be move to, was occupied by another molecule or was located outside the lateral limits of the network domain. The molecules at the entrance could move only in the axial downward direction, being kept for every discrete time $t^*$, a constant value for the concentration at this section ($C_0$, k = 1). At the column exit (k = 30) the concentration (C) was monitored as a function of the discrete time ($t^*$).

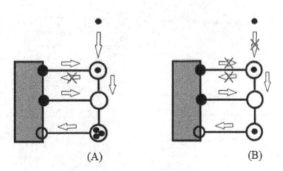

**Figure 5.** Representation of molecule movement along the column: (A) the non-steric model and (B) with steric model. Obs.: The arrows represent the allowed movement that the molecule can do, and crossed arrows represent movement that can't be done.

An important parameter to be taken into account in the stochastic modelling is the number of simulations (N), which indicates the number of times that the same procedure of the simulation is performed using the same control parameters. The increase in the number of simulations (N) leads to a decrease in the dispersion of the calculated value.

The steric effects were also investigated, that is, while the solid phase is able to adsorb one molecule at each site, one possibility is that the liquid phase site is able to contain one molecule and another possibility is that it is able to contain a unlimited number of molecules. In Fig. 5 are represented those two possibilities, being (A) for the case without the steric limitation and (B) representing the steric restriction.

## 3. Results and discussions

### 3.1. Langmuir isotherm model for surface adsorption

In Fig. 6 are represented the adsorption isotherms: in (A) are the results obtained for the simulations using a square network and in (B) are the experimental and deterministic model adapted from Silva (2004). One can observe that the results obtained using the stochastic adsorption method are representative for the studied phenomenon with the classic Langmuir isotherm.

These results show that the stochastic phenomenology is determinant to the behavior of equilibrium systems with multimolecules, and the overall result is governed by the individual actions of each component.

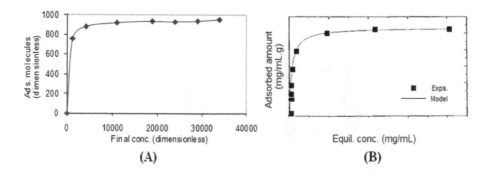

**Figure 6.** Results obtained with the stochastic simulation (A) and the experimental and classical modelling (B) by Silva (2004).

Continuous Chromatography Modelling with 2D and 3D Networks and Stochastic Methods –
Effects of Porous Structure and Solute Population

59

## 3.2. Results for the diffusion phenomenon (2D and 3D modelling)

In Figs. 7, 8 and 9 are represented the percolation evolution both in the square and cubic networks using different moving mechanisms. In Fig. 7 are represented the percolation using a square network of 50x50 nodes, for a time equal to 2000 steps, for four directions, and occupation probability $p_{oc} = 1$ (Fig. 7 A) and $p_{oc} = 0.7$ (Fig. 7 B).

In Fig. 8 are represented results for the same conditions considered before, but allowing only 3 directions for the dispersion mechanism. One can observe a greater tendency towards the axial diffusion.

A evolution percolation method for 3D is presented in Fig. 9 for two movement mechanisms.

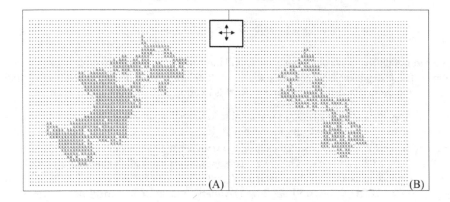

**Figure 7.** Diffusion through the square network for four directions with $p_{oc} = 1$ (A) and $p_{oc} = 0.7$ (B).

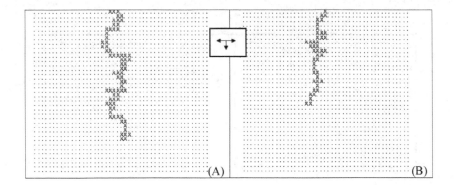

**Figure 8.** Diffusion through the square network for three directions with $p_{oc} = 1$ (A) and $p_{oc} = 0.7$ (B).

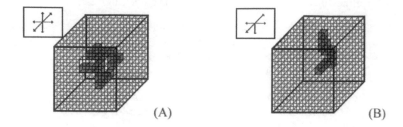

**Figure 9.** Representation for cubic percolation considering six directions (A) and five directions (B).

In Figs. 10 and 11 are presented the distance $(d_p)$ with time in log scale. Using this scale, the power law (Biasse et al., 2010) represented by Eq. (5) becomes

$$\log (d_p) = \log (c) + k.\log (t) \tag{10}$$

We can obtain the expoent $k$, that corresponds to the angular coefficient of the expression above. The $k$ exponent is important because of its relation to the molecular diffusivity and time, being this a fundamental parameter for this relation and sensitive to those values.

From the analysis of Figs. 10 and 11 we are able to observe that diffusivity, which presents a relation with $d_p$, has a direct link to the power law established by Eq. (5), once the model leads to a good correlation with the simulated data for the percolation both for the square (2D) and cubic (3D) networks. The results presented high correlation coefficients, which indicates a good agreement to the power law. Both situations for the regular network with $p_{oc}=1$, for the 2D (Fig. 10A) and 3D (Fig. 11A) networks, presented ideal percolation with universal exponents equal to 0.5014 and 0.4987, respectively. In those cases, because both are regular, with $p_{oc}=1$ and movement in all directions, the correlation coefficient was very closed to 1 $(R^2=1)$.

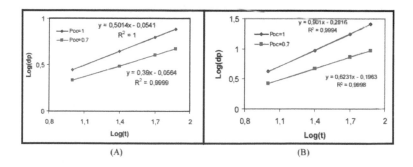

**Figure 10.** Results for 2D networks for four (A) and three (B) moving directions

It is observed, in axial and radial dispersion situations, represented by Figs. 10B and 11B, a significant increase in the value of $k$ exponent when compared to the conditions with diffusion in all directions, demonstrating that diffusion was higher in those cases. For example, we can point out the difference between Figs. 10A and 10B (with $p_{oc}=1$), where the $k$ exponent was increased by 79,7%. With respect to the porosity $\varepsilon$ effect, that has a relation to the $p_{oc}$ was observed that the reduction of $p_{oc}$ from 1 to 0.7 had a more significant effect over the $k$ exponent on 2D networks than on 3D ones. This effect can be observed by the greater slope in Fig. 10. Therefore the porosity reduction leads to a more significant slope on the diffusion for the 2D network represented by lower $k$ exponents. The 3D network, has a lower percolation threshold when compared to the 2D network, presenting a structure with a larger number of possible percolation ways, that with the possibility for the "random walk" to go over larger distances, leading therefore to the larger values of $k$ exponents when compared to 2D networks with the same value of $p_{oc}<1$.

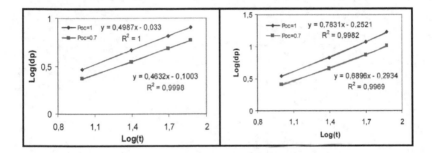

**Figure 11.** Results for 3D network for six (A) and five (B) movement directions.

The analysis of the power law Eq. 5 (Biasse et al., 2010), not only related to the distance $(d_p)$ but mainly in terms of $(A_p)$ and $(V_p)$, took to equivalent results to those previously reported, showing a good agreement of the experimental data with the power law model assumed. Table 1 presents the values of $k$ exponents and the corresponding correlation coefficients. The values for $A_p$ and $V_p$ were obtained in percolated site units, being these units respectively related to 2D and 3D. Such as in the previous cases reported, related to $d_p$, it was observed a reduction of exponent $k$ for the axial and radial dispersion (three and five directions) situations. It was also observed a more significant reduction of $k$ with the porosity reduction in 2D networks.

### 3.3. Results of 3D network column with advective phenomenon

All results presented in this section were obtained for 5000 simulations. In Figs. 12 and 13 the test results varying $p_{ads}$ are presented. One can observe that for $p_{ads}=0$ there is a concentration step, because since no molecule is retained at each time, all of them are allowed to move to the next column section until they reach the exit section. Another observation is that when $p_{ads}$ is increased there is a delay in the variation of the concentration at the column exit, because the chances for the molecule to be adsorbed are higher.

|  | 2D | | 3D | |
|---|---|---|---|---|
|  | 4 directions | 3 directions | 6 directions | 5 directions |
| $P_{oc} = 1$ | 0.7706 ($R^2$=0.9986) | 0.929 ($R^2$=0.9997) | 0.8951 ($R^2$=0.9997) | 0.9304 ($R^2$=0.9997) |
| $P_{oc} = 0.7$ | 0.636 ($R^2$=0.9996) | 0.6413 ($R^2$=0.9998) | 0.8201 ($R^2$=0.9993) | 0.8449 ($R^2$=0.9991) |

**Table 1.** Exponents ($k$) from power law related to the area $A_p$ (2D) and volume $V_d$ (3D).

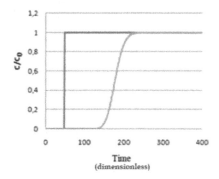

**Figure 12.** Chromatography column exit for $p_{ads}$=0 and $p_{ads}$=1.

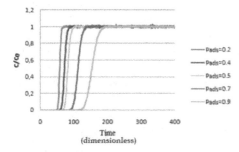

**Figure 13.** Chromatography column exit for different values of $p_{ads}$.

Continuous Chromatography Modelling with 2D and 3D Networks and Stochastic Methods –
Effects of Porous Structure and Solute Population

63

In Fig. 14 it is presented a comparison of the simulation results, without considering axial dispersion, with the experimental data acquired by Cruz (1997). One can observe the good results obtained by the network simulations, despite the breakthrough curve too sharp when equilibrium is reached ($C/C_0 = 1$), in which the curve does not fit well, probably because of the effects of axial dispersion.

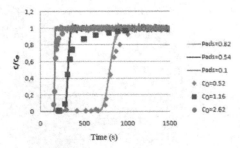

**Figure 14.** Dimensionless concentration at the column exit as a function of time. A comparison between the experimental data of Cruz (1997) and network simulations without axial dispersion

Another important parameter to be considered here is the equilibrium fraction ($k_{eq}$), i.e. the fraction between adsorption and desorption. Another way to find the value for that ratio is comparing the amount of particles adsorbed and the amount of particles in the liquid phase ($q_e/c_e$). Using this model one can confirm that there is a direct relationship between the concentration fraction and the $P_{ads}/P_{des}$. See Table 2.

| Concentration (Simulation) | $k_{eq}$ | $P_{ads}/P_{des}$ |
|---|---|---|
| $C_0 = 0.52$ | 4.455 | 4.555 |
| $C_0 = 1.16$ | 1.222 | 1.173 |
| $C_0 = 1.76$ | 0.466 | 0.470 |
| $C_0 = 2.06$ | 0.353 | 0.369 |
| $C_0 = 2.62$ | 0.160 | 0.111 |

**Table 2.** Relationship between $k_{eq}$ and $P_{ads}/P_{des}$

In Fig. 15 the same network, as the one used to obtain the results shown in Fig. 14 and Table 2, was considered assuming also the axial dispersion phenomenon.

Finally, in Fig. 16 are presented the simulation results (3D network model) for the discrete relative concentration ($C/C_0$) as a function of the adsorption probability ($C/C_0$) taking into account the steric and non-steric effects. The relative concentration indicates a measure of the porous medium resistance to the molecules percolating through the system.

The resistance to the percolation through the chromatographic column decreases as the relative concentration $(p_{ads})$ increases and vice-versa. Therefore, values close to one show a structure without any resistance to the molecules to percolate the column structure.

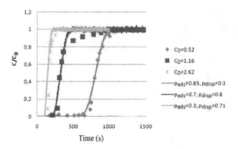

**Figure 15.** Dimensionless Concentration at the column exit as a function of time. Comparison of experimental data acquisition by Cruz (1997) and network simulation considering axial dispersion.

The highest medium resistance corresponds to an adsorption probability near 0.68 which is the minimum point of the curve. At this point the number of molecules percolating the system is reduced to the lowest level.

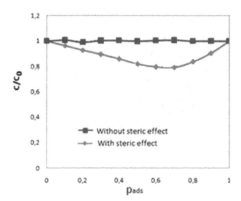

**Figure 16.** Relative concentration as a function of the adsorption probability considering steric and non-steric effects.

It must be noted that the medium resistance disappears at adsorption probability equal to 1 as at this point the adsorption is irreversible (no desorption is observed), and no molecule already adsorbed goes back to the liquid phase of the cavity, and therefore there is no interference in the movement of the other molecules.

## 4. Conclusions

The network stochastic surface multimolecular modelling was able to represent the behavior of Langmuir type adsorption isotherms. Such computational tool can be used to better understand the adsorption mechanisms to surfaces, resulting in the improvement of adsorbent materials.

The diffusion modelling through the 2D and 3D network "random walks" presented results obeying the power law with universal exponents well defined, being the latter related to the diffusive coefficients.

The effects of the power law were observed through the distance $^c/_{c_0}$, the area $d_p$ and the volume $A_p$, being these values related to the diffusion phenomenon. Axial and radial dispersion mechanisms were also represented with a power law modelling, being that behavior related the variations of porosity $V_d$ and probability of occupation $\varepsilon$.

A final important result is the observation of the existence of a direct relationship between the adsorption and desorption probabilities ratio with the equilibrium constant $p_{oc}$.

## Acknowledgements

The authors acknowledge the financial support provided by CNPq, Conselho Nacional de Desenvolvimento Científico e Tecnológico, FAPERJ, Fundação Carlos Chagas Filho de Amparo à Pesquisa do Estado do Rio de Janeiro, and CAPES, Coordenação de Aperfeiçoamento de Pessoal de Nível Superior.

## Author details

Leôncio Diógenes T. Câmara[1], Jader Lugon Junior[2], Flávio de Matos Silva[1],
Guilherme Pereira de Oliveira[1], Lídice Camps Echevarria[3], Orestes Llanes Santiago[3] and
Antônio J. Silva Neto[1]

1 Instituto Politécnico, Universidade do Estado do Rio de Janeiro, Brazil

2 Instituto Federal de Educação, Ciência e Tecnologia Fluminense, Brazil

3 Facultad de Ingeniería Eléctrica - Instituto Superior Politécnico José Antonio Echeverría, Cuba

# References

[1] Biasse, A. D., Fonseca, E. X., Silva, Neto. A. J., & Câmara, L. D. T. (2010). Power-Laws of Diffusion Phenomena In Chromatographic Columns, Advances and Applications in Mathematical Sciences , 2(2), 313-320.

[2] Bird, R. B., Stewart, W. E., & Lightfoot, E. N. (2002). Transport Phenomena, second ed., Wiley.

[3] Bryntesson, L. M. (2002). Pore Network Modelling of the Behaviour of a Solute in Chromatography Media: Transient and Steady-State Diffusion Properties,. J. Chromatogr. A, 945, , 103-115.

[4] Cruz, J. M. (1997). Insulin Adsorption in Ion Exchanger Resins using Fix and Fluidzed Beds, M.Sc. Dissertation, Chemical Engineering Faculty, UNICAMP. (In Portuguese).

[5] Geng, A., & Loh, K. C. (2004). Effects of Adsorption Kinetics and Surface Heterogeneity on Band Spreading in Perfusion Chromatography- A Network Model Analysi. s, Chem. Eng. Sci. 59, , 2447-2456.

[6] Kier, L. B., Cheng, C. K., & Karnes, H. T. (2000). A Cellular Automata Model of Chromatography,. Biomed. Chromatogr. 14, , 530-534.

[7] Loh, K. C., & Geng, A. (2003). Hydrodynamic Dispersion in Perfusion Chromatography- A Network Model Analysis, Chem. Eng. Sci. 58, , 3439-3451.

[8] Loh, K. C., & Wang, D. I. C. (1995). Characterization of Pore Size Distribution of Packing Materials Used in Perfusion Chromatography Using Network Model,. J. Chromatogr. A, 718, , 239-255.

[9] Oliveira, G. P., Câmara, L. D. T., & Silva, Neto. A. J. (2008). Modelling of the Separation Mechanisms in Chromatographic Columns through Adsorption and Advection in Two Dimensional Networks, XI Computational Modelling Meeting, Volta Redonda, Brazil. (In Portuguese)

[10] Silva,(2000). Study of Insulin Adsorption in Ion Exclange Resin Columns: Experimental Parameters and Modelling, D.Sc. Thesis, Chemical Engineering Faculty, UNICAMP. (in Portuguese)

[11] Stauffer, D., & Aharony, A. (1992). Introduction to Percolation Theory. Second Edition.

# Chapter 4

# Minimum Dissipation Conditions of the Mass Transfer and Optimal Separation Sequence Selection for Multicomponent Mixtures I

A.M. Tsirlin and I.N. Grigorevsky

Additional information is available at the end of the chapter

## 1. Introduction

### 1.1. The mass transfer process with minimum irreversibility

In many processes, heat and mass transfer are distributed in time or space. The problem of thermodynamically perfect organization lies in the choice of such concentration and temperature change, in space or time, laws to minimize the entropy production $\sigma$. Below we consider stationary processes and a spatial distribution, for definiteness.

#### 1.1.1. Optimal organization of an irreversible mass transfer process

Consider the irreversible process of mass transfer, in which from one flow to another one substance is transmitted. The problem of minimal irreversibility of this process at a given average intensity of mass transfer takes the form:

$$\sigma = \int_0^L \frac{g(c_1, c_2)}{T}(\mu_1(c_1) - \mu_2(c_2))dl \to \min \tag{1}$$

under conditions

$$\int_0^L g(c_1, c_2)dl = \bar{g}; \tag{2}$$

$$\frac{d(G_1 c_1)}{dl} = \frac{dN_1}{dl} = -g(c_1, c_2); \quad N_1(0) - N_1^0. \tag{3}$$

The minimum is searched by selecting the concentration $c_2(l)$. change law. (In the equation (1) $\mu_i(c_i)$ denotes the chemical potential of the $i$-th flow dependence on the concentration of the redistributed substance in it). Here $G_1$ and $c_1$ is an amount and molarity of the redistributed substance respectively, $N_1$-number of moles of that component.

Minimal irreversibility conditions of mass transfer arise from the solution of (1) — (3). They can be described as follows [24]: *In the mass transfer process with minimum irreversibility the ratio of flow $g$ and chemical potential $\mu_2$ derivatives with respect to the concentration $c_2$ is proportional to the ratio of relative flow $g$ square to the temperature, in any cross-section of $l$.*

$$\frac{\partial g / \partial c_2}{\partial \mu_2 / \partial c_2} = \xi \frac{g^2}{T} \tag{4}$$

Indeed, the entropy production after transition from $dt$ to $dN$ is:

$$\sigma = \int_{N_1^0}^{N_1^L} -\frac{1}{T}(\mu_1(c_1) - \mu_2(c_2)) dN_1 \rightarrow \min \tag{5}$$

under the condition

$$\int_{N_1^0}^{N_1^L} -\frac{1}{g(c_1, c_2)} dN_1 = L. \tag{6}$$

The Lagrange function of problem (5), (6) takes the form

$$R = \frac{1}{T}(\mu_1(c_1) - \mu_2(c_2)) + \lambda \frac{1}{g(c_1, c_2)}, \tag{7}$$

stationarity conditions with respect to $c_2$:

$$\frac{\partial R}{\partial c_2} = 0 \rightarrow \frac{1}{T}\frac{\partial \mu_2}{\partial c_2} + \lambda \frac{1}{g^2(c_1, c_2)}\frac{\partial g}{\partial c_2} = 0 \tag{8}$$

Minimum Dissipation Conditions of the Mass Transfer and Optimal Separation Sequence
Selection for Multicomponent Mixtures I

69

lead to the equation (4). The proportionality coefficient $\xi$ in (4) is defined from the initial data of the current problem.

For a specific task $g(c_1, c_2)$ condition (4) allows us to find a relation between the $c_1$ and $c_2$ for the optimal mass transfer organization. For example, if

$$g = k\left( \frac{\mu_1(c_1)}{T} - \frac{\mu_2(c_2)}{T} \right),$$
(9)

we'll get the following equation from (4) :

$$\mu_1(c_1) - \mu_2(c_2) = const.$$
(10)

At the constant temperature and pressure, this condition leads to the equation

$$c_1(l)/c_2(l) = const$$
(11)

and the constancy of flow $g(c_1, c_2)$ for any $l$.

For the mass transfer law of the form

$$g = k(c_1(l) - c_2(l))$$
(12)

Derivatives are

$$\partial g / \partial c_2 = -K; \quad \frac{\partial \mu_2}{\partial c_2} = \frac{RT}{c_2}.$$
(13)

After their substitution into (4), we obtain

$$(c_1(l) - c_2(l))^2 / c_2(l) = const.$$
(14)

During the mass transfer between phases the driving force of the process is expressed as the difference between the concentration of a redistributed component in one phase $c_1$ and the equilibrium concentration $c_1^P(c_2)$ linearly independent of $c_2$ (concentration of the same component in another phase). In this case, $c_1^P$ is substituted in (11) or (14) instead of concentration $c_2$.

*1.1.2. Example*

Let optimality conditions of irreversible mass transfer have the form (11). From the view of flow $g = \bar{g}/L$ constancy from (3) follows:

$$c_1^*(l) = c_1^0 - \frac{\bar{g}}{rL}; \quad c_1^{p^*}(l) = c_1^0 - \frac{c_1^*(l)}{M},$$ (15)

where M denotes the right side of (11). Substituting $c_1^*$ and $c_1^{p^*}$ in the expression for the mass flux and taking the constancy of this flux into account, we obtain

$$\frac{k}{T}(\mu_1(c_1^*) - \mu_1(c_1^{p^*})) = \frac{\bar{g}}{L},$$ (16)

Or $kR\ln M = \bar{g}/L$ , from which $M = e^{\bar{g}/kLR}$.

Assuming a linear dependence

$$c_1^p(c_2) = ac_2 + b,$$ (17)

where $a, b$ — are some constants determined by processing experimental data of the equilibrium. Then, find the optimum profile

$$c_2^*(l) = \frac{c_1^0 rL - \bar{g}l}{arL} e^{-\bar{g}/kRL} - \frac{b}{a}.$$ (18)

# 2. Irreversible work of separation and heat-driven separation

## 2.1. Introduction

The minimal amount of energy needed for separation a mixture with a given composition can be estimated using reversible thermodynamics. These estimates turn out to be very loose and unrealistic. They also do not take into account kinetic factors (laws and coefficients of heat and mass transfer, productivity of the system, etc.). In this paper we derive irreversible estimates of the work of separation that take into account all these factors.

The majority of separation systems are open systems that exchange mass and energy with the environment. If mass and heat transfer coefficients (determined by the size and construction of the apparatus) are finite and if the productivity of the system is finite then the processes in such systems are reversible. The energy flows, the compositions of the mass flows, and the productivity of the system are linked via the balance equations of energy, mass, and entropy. The latter also includes entropy production in the system. Minimal energy used for separation corresponds to minimal entropy production in the system subject to various constraints. This allows us to estimate this minimal energy.

There is a qualitative as well as a quantitative difference between the reversible and irreversible estimates obtained in this paper. For example, the irreversible estimate of the work of separation for poor mixtures (where the concentration of one of the components is close to one) tends to a finite nonzero limit, which depends on the kinetics factors. The reversible work of separation for such mixtures tends to zero. The reversible estimate differs from the amount of energy needed in practice for separation of poor mixtures by a factor of $10^5$.

For heat-driven separation processes the novel results obtained in this paper include the estimate of the minimal heat consumption as a function of kinetic factors and the thermodynamic limit on the productivity of a heat-driven separation.

## 2.2. Thermodynamic balances of Separation Processes and the Link between Energy Consumption and Entropy Production

Consider the system, shown in Figure 1, where the flow of mixture with rate $g_0$, composition $x_0$, temperature $T_0$, and pressure $P_0$ is separated into two flows with the corresponding parameters $g_i$, $x_i$, $T_i$,$P_i (i=1, 2)$. The flow of heat $q_+$ with the temperature $T_+$ can be supplied, and the flow of heat $q_-$ with the temperature $T_-$ can be removed. The mechanical work with the rate (power) $p$ can be supplied.

In centrifuging, membrane separation, and adsorption–desorption cycles that are driven by pressure variations, no heat is supplied/removed and only mechanical work is spent. In absorption–desorption cycles, distillation, and so forth, no mechanical work is spent, only heat is consumed (heat-driven separation). In some cases the number of input and output flows can be larger. As a rule one can still represent the system as an assembly of separate blocks, whose structure is shown in Figure 1.

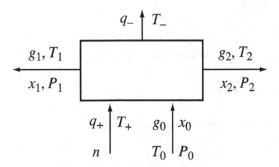

**Figure 1.** Simplified schematic of thermodynamic balances for separation processes.

## 2.2.1. Heat-driven separation

Consider a heat-driven separation ($p=0$) and assume that each of the vectors $x_i=(x_{i1}, ...., x_{ij}, ...., x_{ik})$ $i=0, 1, 2$ consists of $k$ components which denote the molar fraction of the $j$-th substance in the $i$-th flow. The thermodynamic balance equations of mass, energy, and entropy here take the following form

$$g_0 x_{0j} - g_1 x_{1j} - g_2 x_{2j} = 0, \quad j=1,...,k \tag{19}$$

$$\sum_{j=1}^{k} x_{ij} = 1, \quad i=0,1,2 \tag{20}$$

$$q_+ - q_- + g_0 h_0 - g_1 h_1 \quad g_2 h_2 = 0 \tag{21}$$

where $h_i$ is the enthalpy of the i-th flow;

$$\frac{q_+}{T_+} - \frac{q_-}{T_-} + g_0 s_0 - g_1 s_1 - g_2 s_2 + \sigma = 0 \tag{22}$$

$\sigma$ denotes entropy production. From eq (19), eq (20) follows that $g_0 = g_1 + g_2$. After elimination of $g_0$ from eqs (21) and (22) and introduction of enthalpy increments $\Delta h$ and entropy increment $\Delta s$ we get

$$q_+ - q_- + g_1 \Delta h_{01} + g_2 \Delta h_{02} = 0 \tag{23}$$

$$g_2 \Delta s_{02} + g_1 \Delta s_{01} + \frac{q_+}{T_+} - \frac{q_-}{T_-} + \sigma = 0 \tag{24}$$

Here, $\Delta h_{0i} = h_0 - h_i$, $\Delta s_{0i} = s_0 - s_i$ ($i=1, 2$).

Elimination of $q_-$ using eq (23) and its substitution into eq (24) yields

$$\sum_{i=1}^{2} g_i \left( \Delta s_{0i} - \frac{\Delta h_{0i}}{T_-} \right) + q_+ \left( \frac{1}{T_+} - \frac{1}{T_-} \right) + \sigma = 0$$

and the flow of used heat for heat-driven separation is

$$q_+ = \frac{T_+}{T_+ - T_-} \left[ \sum_{i=1}^{2} g_i (\Delta s_{0i} T_- - \Delta h_{0i}) + \sigma T_- \right] \tag{25}$$

The first term in the square brackets depends only on the parameters of the input and output flows and represents the reversible work of separation per unit of time (reversible power of separation). The second term there represents the process kinetics and corresponding energy dissipation.

For mixtures that are close to ideal gases and ideal solutions, molar enthalpies and entropies $h_i$ and $s_i$ in the eqs (21) and (22) can be expressed in terms of compositions and specific enthalpies and entropies of the pure substances. We obtain for each of the flows

$$\Delta h_{0i} = \sum_{j=1}^{k}\left[ x_{0j}h_j(T_0,P_0) - x_{ij}h_j(T_i,P_i) \right]$$

$$\Delta s_{0i} = \sum_{j=1}^{k}\left[ x_{0j}s_j^0(T_0,P_0) - x_{ij}s_j^0(T_i,P_i) - R(x_{0j}\ln x_{0j} - x_{ij}\ln x_{ij}) \right], \quad i = 1,2 \tag{26}$$

where $R$ is the universal gas constant. The reversible energy consumption here is

$$q_+^0 = \frac{1}{\eta_k}\sum_{i=1}^{2}g_i\sum_{j=1}^{k}\left[ [x_{0j}s_j^0(T_0,P_0) - x_{ij}s_j^0(T_i,P_i) - R(x_{0j}\ln x_{0j} - x_{ij}\ln x_{ij})]T_- + x_{ij}h_j(T_i,P_i) - x_{0j}h_j(T_0,P_0) \right] \tag{27}$$

We denote here the Carnot efficiency of the ideal cycle of the heat engine as

$$\eta_C = \frac{T_+ - T_-}{T_+}$$

Condition (25) can be rewritten as

$$q_+ = \frac{1}{\eta_C}(p^0 + \sigma T_-) \tag{28}$$

Here, $p_0$ is the reversible power of separation that is equal to the reversible flow of heat given by eq (27) multiplied by the Carnot efficiency. When eq (28) was derived we took into account only the irreversibility $\sigma$ of the separation process (the irreversibility of the heat transfer was not taken into account). In reality heat can be supplied/removed with a finite rate only irreversibly. Any transformation of heat into work with finite heat transfer coefficients and finite power is irreversible. This leads to a lower efficiency than the Carnot efficiency. The closed form expression for this efficiency was obtained in ref [16]. It depends on the power p and on heat transfer coefficients for heat supply and heat removal $\alpha_+$ and $\alpha_-$. For the Newton (linear) law of heat transfer it has the form

$$\eta_p = \max\frac{p}{q_+} = 1 - \frac{1}{2T_+}\left( T_+ + T_- - \frac{4p}{\alpha} - \sqrt{(T_+ - T_-)^2 + \left(\frac{4p}{\alpha}\right)^2 - 8\frac{p}{\alpha}(T_+ + T_-)} \right) \tag{29}$$

where it is assumed that there is constant contact of the working body with the heat reservoirs and

$$\alpha = \frac{4\alpha_+\alpha_-}{\alpha_+ + \alpha_-} \tag{30}$$

It is easy to show that if $p \to 0$ then $\eta_p$ tends to the Carnot efficiency.

Substitution of $\eta_p$ instead of $\eta_C$ in eq (28) allows us to derive a tighter estimate for the heat consumption in heat-driven separation processes by finding the minimal possible entropy production $\sigma$ subject to various constraints

$$q_+ \geq q_+^{\min} = \frac{p^{\min}}{\eta_p(p^{\min},\alpha,T_+,T_-)} \tag{31}$$

where

$$p^{\min} = p^0 + \sigma^{\min}T_- \tag{32}$$

Conditions (29-31) single out the area of thermodynamically feasible heat-driven separation systems.

Expressions (27) and (28) and eq (25) can be further specified by assuming the constancy of heat capacities, that the mixture is binary, and so forth.

### 2.2.2. Mechanical separation

Consider a separation system that uses mechanical work with rate $p$. Assume that no heat is supplied/removed ($q_+=q_-=0$) and that input and output flows have the same temperature $T$ and the same pressure. Multiplication of eq (24) by $T$ and subtraction of the result from the energy balance eq (23), where ($q_+-q_-$) is replaced with the supplied power $p$, yields

$$p = T\sigma + g_0\sum_{i=1}^{2}\gamma_i(T\Delta s_{0i} - \Delta h_{0i}) \tag{33}$$

here $\gamma_i = g_i/g_0$

After taking into account eq (27) that the enthalpy increment $\Delta h_{0i}$ in a mechanical separation is zero, we get

$$p = g_0 RT \left[ \sum_{i=1}^{2} \gamma_i \sum_{j=1}^{k} x_{ij} \ln x_{ij} - \sum_{j=1}^{k} x_{0j} \ln x_{0j} \right] + T\sigma = p^0 + T\sigma \qquad (34)$$

The first term in this expression represents the minimal power for separation that corresponds to the reversible process ($\sigma = 0$). This power $p^0$ is equal to the difference between the reversible power for complete separation of the input flow $p_0^0 = -g_0 RT \sum_j x_{0j} \ln x_{0j}$ and the combined reversible power of separation of the output flows $p_1^0$ and $p_2^0$.

Here

$$p_i^0(x_i) = -RT g_0 \gamma_i \sum_{j=1}^{k} x_{ij} \ln x_{ij}, \quad i = 0,1,2 \qquad (35)$$

is the reversible power of separation of the $i$-th flow into pure substances.

## 2.3. Minimal work of separation in irreversible processes

### 2.3.1. Assumptions and problem formulation

Assume that the components of the input mixture are close to ideal gases or ideal solutions. The chemical potential of the $i$-th component can then be written in the following form

$$\mu_i(T,P) = \mu_0(T,P) + RT \ln x_i, \quad i = 1,...,k \qquad (36)$$

where $x_i$ is the concentration of the $i$-th component.

First we consider a system that includes three elements, a reservoir with the time independent temperature $T$, pressure $P$, and vector of concentrations $x_0 = \{x_{01}, ..., x_{0k}\}$ (therefore its chemical potential $\mu_0$ is also time independent), the finite capacity output subsystem with chemical potential $\mu_1$ that depends on the current compositions of the mixture and of the working body that has controllable values of chemical potential $\mu_o^w$ and $\mu_1^w$, at the points of contact with reservoir and output subsystem. At the time the intensive variables of the output subsystem coincide with the values of the reservoir's intensive variables, and the number of moles in it is given and equal to $N_0$. At time $\tau$ the number of moles $N(\tau)$ and the composition $x(\tau)$ in the output subsystem are given. The mass transfer coefficients between the reservoir and the working body and the working body and the output subsystem are finite and fixed. The minimal necessary work required for the separation is sought.

We do not consider here how to implement the derived optimal dependence of the chemical potential of the working body because of two reasons. First, our main objective is to derive a lower bound on the work of separation. However, imposing constraints on feasible variations of chemical potential would lead to an increase in energy consumption. Second, we will demonstrate that for the majority of mass transfer laws the optimal mass transfer flow is time independent, and its implementation is straightforward.

The work of separation in an isothermal process for an adiabatically insulated system can be found using the Stodola formula in terms of the reversible work $A_0$ and the entropy increment $\Delta S$

$$A = A_0 + T\Delta S \tag{37}$$

The reversible work is equal to the increment of the system's internal energy. Since as a result of the process $(N(\tau) - N(0))$ moles of mixture with the composition $x_0$ is removed from the reservoir, and the energy of the output subsystem rises because of the increase of the amount of moles in it from $N(0)$ to $N(\tau)$ and its composition from $x_0$ to $x_\tau$, the total change of the system's internal energy is

$$A_0 = N(\tau)\sum_{i=1}^{k}\Delta\mu_i = N(\tau)RT\sum_{i=1}^{k}[x_i(\tau)\ln x_i(\tau) - x_{i0}\ln x_{i0}] \tag{38}$$

and it is independent of $N(0)$. Because $A_0$ is determined by $N$, $x(\tau)$, $x(0)$, the minimum of $A$ corresponds to the minimum of the entropy increment

$$\begin{aligned}\Delta S &= \frac{1}{T}\int_0^\tau \sum_{i=1}^{k}[g_{0i}(\mu_{0i}-\mu_i^w) + g_{1i}(\mu_i^w - \mu_{1i})]dt \\ &= \frac{1}{T}\int_0^\tau \sum_{i=1}^{k}(g_{0i}\Delta\mu_{0i} + g_{1i}\Delta\mu_{1i})dt\end{aligned} \tag{39}$$

Because the working body's parameters have the same values at the beginning and at the end of a cycle

$$\int_0^\tau g_{i0}dt = \int_0^\tau g_{i1}dt$$
$$N(\tau)x_i(\tau) - N(0)x_i(0) = \Delta(Nx_i), \quad i = 1,2,...,k \tag{40}$$

### 2.3.2. Optimal solution

The problem of minimization of $\Delta S$ subject to constraints eq (40) on $g_{0i} \geq 0$, $g_{1i} \geq 0$ becomes simpler in a common case where the chemical potentials' increments $\Delta\mu_{0i}$, $\Delta\mu_{1i}$are unique functions of flows $g_{0i}$ and $g_{1i}$, correspondingly. If processes are close to equilibrium then this dependence is linear.

Assume

$$\Delta \mu_{0i} = \phi_{0i}(g_{0i}), \quad \Delta \mu_{1i} = \phi_{1i}(g_{1i})$$

then the problems (39) and (40) can be decomposed into 2k problems

$$\Delta S_{ji} = \int_0^\tau \sigma_{ji}(g_{ji})dt \rightarrow \min \Big/ \int_0^\tau g_{ji}dt = \Delta(Nx_i) \quad j=0,1, \quad i=1,2,...,k \qquad (41)$$

where $\sigma_{ji} = g_{ji}\phi_{ji}(g_{ji})$ is the function that determines dissipation.

Problems eq (41) are averaged nonlinear programming problems. Their optimal solutions $g_{ji}^*$ are either constants and equal to

$$g_{ji}^* = g_{1i}^* = \frac{\Delta(Nx_i)}{\tau} \qquad (42)$$

or switches between two so-called basic values on the interval $(0, \tau)$, the solution eq (42) corresponding to the case where the convex envelope of the function $\sigma_{ji}(g_{ji})$ is lower than the value of this function at $g_{ji}^*$. Characteristic forms of the function $\sigma_{ji}(g_{ji})$ for the constant and switching regimes are shown in Figure 2.

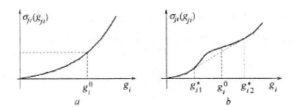

**Figure 2.** Dependence of the entropy production on the rate for the constant (a) and switching (b) solutions ($g_{1i}^*$ and $g_{2i}^*$ are the basic values of the rate).

If the function $\sigma_{ji}$ is concave then the optimal rate $g_{ji}$ is always constant. Let us calculate the second derivative of $\sigma$ on $g$ (we omit subscripts for simplicity). If it is positive then the constancy of the rate in the optimal process is guaranteed.

$$\sigma''(g) = 2\phi'(g) + g\phi''(g) \geq 0 \qquad (43)$$

The first term in this expression is always positive because the chemical potentials' difference is the driving force of mass transfer and monotonically depends on the flow. For the majority of laws of mass transfer the inequality eq (43) holds. In particular, it holds if the flow of mass transfer is proportional to the difference of chemical potentials in any positive degree.

Consider mass transfer flow that depends linearly on the chemical potential difference for all $i, j$. Then

$$g_{ji} = \alpha_{ji} \Delta \mu_{ji} \rightarrow \phi_{ji} = \frac{g_{ji}}{\alpha_{ji}} \tag{44}$$

It is clear that the conditions eq (43) hold and the optimal rates of flows obey equalities (42).

Equalities (42) hold for any nonswitching solution. The minimal increment of the entropy production for such solution is

$$\Delta S^{\min} = \sum_{ij} \Delta S_{ji}^{\min} = \tau \sum_{ij} \sigma_{ji} \left( \frac{\Delta(Nx_i)}{\tau} \right) \tag{45}$$

and the minimal work of separation is

$$A_{\min} = A_0 + \tau T \sum_{ij} \sigma_{ji} \left( \frac{\Delta(Nx_i)}{\tau} \right) \tag{46}$$

The optimal rates are determined by the initial and final states which allows us to specify the estimate eq (46).

Near equilibrium the flows obey Onsanger's kinetics eq (44), and from eq (46) it follows that

$$A_{\min} = A_0 + \tau \sum_{i=1}^{k} g_i^2 \left( \frac{1}{\alpha_{0i}} + \frac{1}{\alpha_{1i}} \right) = A_0 + \frac{1}{\tau} \sum_{i=1}^{k} \frac{\Delta^2(Nx_i)}{\bar{\alpha}_i} \tag{47}$$

$$\bar{\alpha}_i = \frac{\alpha_{0i}\alpha_{1i}}{\alpha_{0i} + \alpha_{1i}} \tag{48}$$

is the equivalent mass transfer coefficient on the i-th component and the minimal entropy production is

Minimum Dissipation Conditions of the Mass Transfer and Optimal Separation Sequence
Selection for Multicomponent Mixtures I

79

$$\sigma_{min} = \frac{1}{T\tau^2} \sum_{i=1}^{k} \frac{\Delta^2(Nx_i)}{\bar{\alpha}_i} \tag{49}$$

The lower bound for the average power of separation is

$$p_{min} = \frac{A_{min}}{\tau} = \frac{A_0}{\tau} + \frac{1}{\tau^2} \sum_{i=1}^{k} \frac{\Delta^2(Nx_i)}{\bar{\alpha}_i} \tag{50}$$

$p_0 = A_0/\tau$ is the reversible power of separation.

If

$$N(0)=0, \quad \Delta(Nx_i)=Nx_i(\tau)$$

then expressions (47) and (50) take the form

$$A_{min} = A_0 + \frac{N^2}{\tau} \sum_{i=1}^{k} \frac{x_i^2(\tau)}{\bar{\alpha}_i} \tag{51}$$

$$p_{min} = p_0 + g^2 \sum_{i=1}^{k} \frac{x_i^2(\tau)}{\bar{\alpha}_i} \tag{52}$$

Where

$$A_0 = NRT \sum_{i=1}^{k} [x_i(\tau)\ln x_i(\tau) - x_i \ln x_i] \tag{53}$$

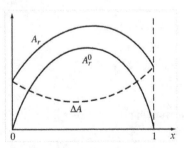

**Figure 3.** Reversible ($A_0$) and irreversible ($A_r$) estimates of the minimal work of separation of binary mixture as functions of key component's concentrations.

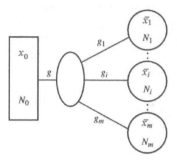

**Figure 4.** Separation of the system with finite capacity on m subsystems.

Note that the irreversible estimate of the work of separation eq (51) does not tend to zero for poor mixtures when the concentration of one of the components tends to one (Figure 3).

If system includes not one but a number of output subsystems then it is clear that the estimate for the minimal work of separation is equal to the sum of the estimates for each subsystem.

$$A_{\min} = \sum_{j=1} A_{\min}^j, \quad p_{\min} = \sum_{j=1} p_{\min}^j \tag{54}$$

The superscript j here denotes the subsystems.

### 2.3.3. Separation of a System with finite capacity into m subsystems

Consider a system that is shown in Figure 4. Its initial state is described by the vector of concentrations $x_0$, the number of moles of the mixture $N_0$, and its final state by the number of moles $N_j$, $j=1, ..., m$ in each of the subsystems and their concentrations, $x_j$. The mass balances yields

$$\sum_{j=1}^{m} \bar{N}_j = N_0$$

$$\sum_{j=1}^{m} \bar{N}_j \bar{x}_{ji} = N_0 x_{0i}, \quad i=1,2,...,k \tag{55}$$

The work in the reversible separation process here is

$$A_r^0(x_0,\bar{x}) = RT \left[ \sum_{j=1}^{m} \bar{N}_j \sum_i \bar{x}_{ji} \ln \bar{x}_{ji} - N_0 \sum_i x_{0i} \ln x_{0i} \right] = A_{r0}^0(x_0, N_0) - \sum_{j=1}^{m} A_{rj}^0(\bar{x}_j, \bar{N}_j) \tag{56}$$

The reversible work of separation is equal to the difference of the reversible work of separation of the initial mixture into pure components and the reversible work of separation for mixtures in each of the subsystems.

We again assume that flows $g_j$ have components $g_{ji}$ proportional to the difference of the chemical potential of the subsystem and the working body with the coefficient $\alpha_{ji}$. Here, the condition of minimal work of separation corresponds to the condition of flow constancy

$$g_{ji} = \frac{\overline{N}_j \overline{x}_{ji}}{\tau}, \quad i = 1, 2, \dots, k, \quad j = 1, \dots, m \tag{57}$$

$$\Delta \mu_{ji} = \frac{g_{ji}}{\overline{\alpha}_{ji}}, \quad j = 0, 1, \dots, m \tag{58}$$

Here, $\overline{\alpha}_{ji}$ is the equivalent mass transfer coefficient calculated using eq (48) for the flow into the j-th output subsystem of the i-th component. Similarly as was done above for the system with the reservoir and one finite capacity output subsystem and flows proportional to the final concentrations eq (57), these concentrations in the output subsystems are time independent and equal to $\overline{x}_j$, correspondingly, and the number of moles $\overline{N}_j(t)$ depends linearly on time. The power p here is constant

$$p = \frac{RT}{\tau} \sum_{j=1}^{m} \overline{N}_j \sum_i \overline{x}_{ji} \ln \frac{\overline{x}_{ji}}{x_{0i}} + \frac{1}{\tau^2} \sum_{j=1}^{m} \overline{N}_j^2 \sum_i \overline{x}_{ji}^2 / \overline{\alpha}_{ji} \tag{59}$$

The minimal work of separation for the mixture with concentrations $x_0$ into m subsystems with concentrations $\overline{x}_i$ over the time $\tau$ is

$$A_r = RTN_0 \sum_{j=1}^{m} \gamma_j \sum_i \overline{x}_{ji} \ln \frac{\overline{x}_{ji}}{x_{0i}} + \frac{N_0^2}{\tau} \sum_{j=1}^{m} \gamma_j^2 \sum_i \overline{x}_{ji}^2 / \overline{\alpha}_{ji} \tag{60}$$

Here, $\gamma_j = N_j / N_0$, $\overline{\alpha}_{ji} = \alpha_{ji} \alpha_{0i} / (\alpha_{0i} + \alpha_{ji})$

The first term here coincides with the reversible work of separation $A_r^0$ of the mixture of $N_0$ moles with concentration $x_0$ into subsystems with number of moles $\overline{N}_j$ and concentrations $\overline{x}_j$. The second term takes into account irreversibility of the process. $A_r$ decreases monotonically and tends to $A_r^0$ when process duration $\tau$ and mass transfer coefficient $\overline{\alpha}_{ji}$ increases.

## 2.3.4. Example

Consider separation of the binary mixture into pure components in time $\tau$. In this case $N_1 = x_0 N_0$, $N_2 = (1 - x_0) N_0$, where $x_0$ is the concentration of the key component, $\bar{x}_{11} = \bar{x}_{22} = 1$. From the formula (60) we get

$$A_r = -RTN_0(x_0 \ln x_0 + (1 - x_0) \ln(1 - x_0)) + \frac{N_0^2}{\tau}\left(\frac{x_0^2}{\bar{\alpha}_{11}} + \frac{(1-x_0)^2}{\bar{\alpha}_{22}}\right) = A_r^0(x_0) + \frac{N_0^2}{\tau}\left(\frac{x_0^2}{\bar{\alpha}_{11}} + \frac{(1-x_0)^2}{\bar{\alpha}_{22}}\right) \qquad (61)$$

The estimate eq (61) was derived in ref [1] by solving the problem of optimal separation of the binary mixture in the given time $\tau$ in Van't Hoff's thought experiment with movable pistons and semitransparent membrane where $\bar{\alpha}_{11}$ and $\bar{\alpha}_{22}$ are the permeability coefficients on the first and second component. If flows do not depend explicitly on the chemical potentials' differentials, for example, are proportional to the concentrations' differential, then an estimate similar to the one obtained above can be constructed by solving the following auxiliary nonlinear programming problem

$$\Delta\mu_i(P_0^i, P_i) \rightarrow \min_{P_0^i, P_i} / \, g_i(P_0^i, P_i) = g_i, \quad i = 1, 2, \dots \qquad (62)$$

Here, $(P_0^i, P_i)$ are partial pressures of the components in contacting subsystems that depend on the chemical potentials' differentials $\Delta\mu_i$. The flow $g_i$ depends on the same differentials. Minimums in these problems are sought for different values of constant $g_i > 0$ and nonpositive $P_0^i$ and $P_i$ We denote the minimal values of the objective in each of these problems $\Delta\mu_i^{\min}(g_i)$ as $\Delta\mu_i^*(g_i)$. This dependence can be used in the estimate eq (41) of the irreversible work of separation.

## 2.3.5. Example

Assume $\Delta\mu = RT \ln(P_0 / P)$, $g(P_0, P) = (P_0 - P)/\alpha$, and $0 < P < P_{\max}$. Let us express $P_0$ in terms of $g$ and $P$:

$P_{0i} = \alpha_i g_i + P_i, \quad i = 1, 2$

$\Delta\mu = RT \ln(\alpha g / P + 1)$ attains its minimum at $P = P_{\max} \, \forall \, g$.

Therefore, $\Delta\mu_i^*(g_i) = RT \ln(\alpha_i g_i / P_{\max} + 1)$.

## 2.4. Potential application of obtained estimates

We will illustrate the possibilities of the application of the derived estimates.

### 2.4.1. *Estimate of the power of separation in a continuous separation system*

Consider a continuous separation system with the input flow $g_0$ with concentration $x_0$ and $m$ output flows $g_j (j=1, ..., m)$ with concentrations $x_j = \{x_{j0}, x_{j1}, ..., x_{jk}\}$. Here, the temperatures on the input and output flows are close to each other.

Equation (59) allows us to estimate the minimal power required for continuous separation in such system

$$p_{\min} = \sum_{j=1}^{m} p_{0j} + g_0^2 \sum_{j=1}^{m} \gamma_j^2 \sum_{i=1}^{k} \frac{x_{ji}^2}{\alpha_{ji}} \tag{63}$$

Where

$$\gamma_j = \frac{g_j}{g_0} \geq 0, \quad \sum_{j=1}^{m} \gamma_j = 1 \tag{64}$$

$$p_{0j} = g_0 \gamma_j RT \sum_{i=1}^{k} [x_{ji} \ln x_{ji} - x_{0i} \ln x_{0i}] p_{0j} = \gamma_j M_j (g_0, x_j) \tag{65}$$

Mass balance equations yield

$$\sum_{j=1}^{m} \gamma_j x_{ji} = x_{0i}, \quad i = 1, ..., k-1,$$

$$\sum_{i=1}^{k} x_{ji} = 1, \quad j = 0, ..., m \tag{66}$$

The number of conditions eq (66) is $k-1$, because the concentration of one of the components is determined by the conditions eq (64).

If the number of flows $m > k$, and their compositions are given, then the removal fractions can be chosen in such a way that the power of separation is minimal subject to constraints eqs (64) and (66). The Lagrange function of this problem is

$$L = \sum_{j=1}^{m} \left\{ \gamma_j M_j + \gamma_j^2 r_j - \lambda_0 \gamma_j - \sum_{i=1}^{k} \lambda_i \gamma_j x_{ji} \right\} \tag{67}$$

here

$$r_j(g_0, x_j) = g_0^2 \sum_{i=1}^{k} \frac{x_{ji}}{\alpha_{ji}}$$

L is the concave function on $\gamma_j$, and its conditions of stationarity determine the flows that minimize the power for separation for a given flow's compositions

$$\gamma_j^* = \frac{\lambda_0 - M_j + \sum_{i=1}^{k-1} \lambda_i x_{ji}}{2r_j}, \quad j = 1, ..., m \tag{68}$$

We have k linear equations for $\lambda_0$ and $\lambda_i$

$$\frac{1}{2} \left[ \sum_{j=1}^{m} \frac{\lambda_0 - M_j}{r_j} + \sum_{i=1}^{k-1} \lambda_i \sum_{j=1}^{m} \frac{x_{ji}}{r_j} \right] = 1 \tag{69}$$

$$\frac{1}{2} \left[ \sum_{j=1}^{m} x_{ji} \left( \frac{\lambda_0 - M_j}{r_j} + \frac{1}{r_j} \sum_{i=1}^{k-1} \lambda_i x_{ji} \right) \right] = x_{0i}, \quad i = 1, ..., k-1 \tag{70}$$

## 2.4.2. Example

Assume $m = 3$, $k = 2$, $g_0 = 1$ mol/s, $T = 300$K, and the compositions and transfer coefficients are

$$x_{01} = x_{02} = 0.5$$

$x_{11} = 0.9$; $x_{12} = 0.1$; $\bar{\alpha}_{11} = \bar{\alpha}_{12} = 0.004$ $mol^2/(J\ s)$

$x_{21} = 0.3$; $x_{22} = 0.7$; $\bar{\alpha}_{21} = \bar{\alpha}_{22} = 0.01$ $mol^2/(J\ s)$

$x_{31} = 0.1$; $x_{32} = 0.9$; $\bar{\alpha}_{31} = \bar{\alpha}_{32} = 0.06$ $mol^2/(J\ s)$

From eq (65) we obtain $M_1 = 910$, $M_2 = 197$, $M_3 = 910$, and $r_1 = 205$, $r_2 = 580$, $r_3 = 137$.

Equations (69) and (70) for $\lambda$-multipliers take the form

$$\frac{1}{2} \left[ \frac{\lambda_0 - M_1}{r_1} + \frac{\lambda_0 - M_2}{r_2} + \frac{\lambda_0 - M_3}{r_3} + \lambda \left( \frac{x_{11}}{r_1} + \frac{x_{21}}{r_2} + \frac{x_{31}}{r_3} \right) \right] = 1$$

$$\frac{1}{2} \left[ x_{11} \left( \frac{\lambda_0 - M_1}{r_1} + \frac{\lambda_1 x_{11}}{r_1} \right) + x_{21} \left( \frac{\lambda_0 - M_2}{r_2} + \frac{\lambda_1 x_{21}}{r_2} \right) + x_{31} \left( \frac{\lambda_0 - M_3}{r_3} + \frac{\lambda_1 x_{31}}{r_3} \right) \right] = x_{01}$$

We obtain $\lambda_0 = 894$, $\lambda_1 = 183$. Their substitution in eq (68) yields $\gamma_1^* = 0.36$, $\gamma_2^* = 0.64$, $\gamma_3^* = 0$ and the corresponding estimate for the minimal irreversible power of separation eq (63) is

Minimum Dissipation Conditions of the Mass Transfer and Optimal Separation Sequence
Selection for Multicomponent Mixtures I

85

$p_{min} = 718$ wt

### 2.4.3. The selection of the separation sequence for a multicomponent mixture

In practice, separation of multicomponent mixtures is often realized via a sequence of binary separations. So, a three-component mixture is first separated into two flows, one of which does not contain one of the components. The second flow is then separated into two unicomponent flows. The reversible work of separation (that corresponds to the power $p_0$) does not depend on the sequence of separation, because $p_0$ is determined by the rates and compositions of the input and output flows of the system as a whole. The irreversible component of the power $\Delta p$ in eq (63) depends on the sequence of separation and can be used to find the optimal one.

Consider a three-component mixture with concentration $x_0 = (x_{01}, x_{02}, x_{03})$, and rate $g_0$ we set to one. We denote the mass transfer coefficients at the first and second stages of separation as $\alpha_1$ and $\alpha_2$. They depend on the construction of the apparatus. First, we assume for simplicity that these coefficients do not depend on the mixture's composition (in the general case they do depend on it). We consider irreversible power consumption for two cases:

**a.** The first component is first separated, then the second and the third are separated.

**b.** The second component is separated, and then the first and the third are separated.

We assume that the separation at each stage is complete. We get up to the constant multiplier

$$\Delta p_a = \Delta p_{a1} + \Delta p_{a2} = x_{01}^2 / \alpha_1 + \frac{(x_{02} + x_{03})^2}{\alpha_1} + (x_{02} + x_{03})^2 + (x_{02}^2 / \alpha_2 + x_{03}^2 / \alpha_3) \qquad (71)$$

The first two terms in this sum represent the loss of irreversibility during the first stage of separation. For $g_0 = 1$ and complete separation the output rates of this stage $g_1$ and $g_2$ are $x_{01}$ and $(x_{02} + x_{03})$, correspondingly.

Consider the first stage of case a for $g_0 = 1$ and complete separation and view the second and third component as the same substance with the output rate $x_{02} + x_{03} = 1 - x_{01}$. The irreversible expenses eq (63) are

$$\Delta p_{a1} = \frac{x_{01}^2}{\alpha_1} + \frac{(1 - x_{01})^2}{\alpha_1} = \frac{2x_{01}^2 + 1 - 2x_{01}}{\alpha_1} \qquad (72)$$

When the second flow is separated into two flows their rates are

$$g_{22} = \frac{x_{02}}{(1 - x_{01})}, \quad g_{23} = \frac{x_{03}}{(1 - x_{01})}$$

and the irreversible power is

$$\Delta p_{a2} = \frac{1}{a_2(1-x_{01})^2}(x_{02}^2 + x_{03}^2)$$

The combined irreversible power is

$$\Delta p_a(x_{01}, x_{02}) = \frac{2x_{01}^2 - 2x_{01} + 1}{a_1} + \frac{x_{02}^2 + (1-x_{01}-x_{02})^2}{a_2(1-x_{01})^2}$$

Similarly in case b we get

$$\Delta p_b(x_{01}, x_{02}) = \frac{2x_{02}^2 - 2x_{02} + 1}{a_1} + \frac{x_{01}^2 + (1-x_{01}-x_{02})^2}{a_2(1-x_{02})^2}$$

The differential between these two values is

$$\Delta p_{ab} = \Delta p_a + \Delta p_b = \frac{2}{a_1}\left[(x_{01}^2 - x_{02}^2) - (x_{01} - x_{02})\right] + \frac{1}{a_2(1-x_{01})(1-x_{02})}\left[(1-x_{02})^2(x_{02}^2 + x_{03}^2) - (1-x_{01})^2(x_{01}^2 + x_{03}^2)\right] \quad (73)$$

If $\Delta p_{ab} > 0$, then sequence b is preferable.

Note that it is not possible to formulate the general rule to choose the optimal separation sequence for a multicomponent mixture, in particular, on the basis of the reversible work of separation. It is necessary here to compare irreversible losses for each sequence.

*2.4.4. Example*

Assume that the composition of the input three component mixture is $x_{01} = 0.6$, $x_{02} = 0.3$, $x_{03} = 1 - x_{01} - x_{02}$; the mass transfer coefficients are $a_1 = 0.01 \text{mol}^2/(\text{J s})$, $a_2 = 0.02 \text{mol}^2/(\text{J s})$. From (eq 73) we find that the difference in power between sequences *a* and *b* is

$$\Delta p_{ab} = \Delta p_a - \Delta p_b = -7.82 \text{ J}$$

The comparison of the combined minimal irreversible power for the same initial data shows that the power for separation of a mixture using sequence *b* is higher than the power used for sequence *a*, that is, $\Delta p_{ab} < 0$.

Thus, sequence *a* is preferable, and it is better to perform the complete separation by separating the first component.

## 2.5. Limiting productivity and minimal heat consumption for a heat-driven separation

In many separation processes a heat engine is used to create the differential of the chemical potential between the working body and the reservoirs (the driving force of mass transfer). Here, the working body is heated during contact with one reservoir and is cooled during contact with the other reservoir. One can represent the heat-driven separation system as a transformer of heat into the work of separation that generates power p, consumes heat flow from hot reservoir $g_+$, and rejects flow $g_-$ to the cold reservoir. Heat transfer coefficients for contacts with the hot and cold reservoir $\alpha_+$ and $\alpha_-$ are fixed.

It was shown in refs [12] and [6] that the potential of the direct transformation of heat to work is limited and the maximal generated power for the working body with the distributed parameters is

$$p_{max} - \bar{\alpha}(\sqrt{T_+} - \sqrt{T_-})^2 \tag{74}$$

In this expression $\bar{\alpha} = (\alpha_+\alpha_-)/(\alpha_+ + \alpha_-)$ is the equivalent heat transfer coefficient for continuous contact with the reservoirs; $\bar{\alpha} = (\alpha_+\alpha_-)/(\sqrt{\alpha_+} + \sqrt{\alpha_-})^2$ is the equivalent heat transfer coefficient for sequential contact.

The maximal power determines the heat flow consumed from the hot reservoir. Further increase of heat consumption for given values of heat transfer coefficients requires an increase of the temperature differential between the reservoirs and the working body and reduces the power.

The dependence of the used power on the productivity of irreversible separation processes is monotonic eq (63). Therefore, the limiting productivity of heat-driven separation processes corresponds to the maximal possible power produced by transformation of heat into work. Further increase of heat consumption $q_+$ reduces power and therefore reduces the productivity of separation process.

For the Newton (linear) law of mass transfer and heat–work transformer the dependence of the power on the heat used is

$$q^+(p) = \frac{p}{\eta_p} = \frac{2p}{\left(\dfrac{p}{\bar{\alpha}T_+} + \eta_C\right) + \sqrt{\left(\dfrac{p}{\bar{\alpha}T_+} + \eta_C\right)^2 - \dfrac{4p}{\bar{\alpha}T_+}}} \tag{75}$$

Here, $\eta_C = (T_+ - T_-)/T_+$ is the Carnot efficiency, $T_+$ and $T_-$ are the hot and cold reservoir's temperatures, and $\bar{\alpha} = (\alpha_+\alpha_-)/(\alpha_+ + \alpha_-)$ is the equivalent heat transfer coefficient.

The minimal heat consumption $q_+$ as a function of productivity $g_0$ for a heat-driven separation can be obtained by substituting expression (75) instead of p in the right-hand side of eq (63). The result holds for $p \le p_{max}$ and therefore for $g_0 \le g_{0max}$. The duration here must not exceed the maximal possible duration.

Substitution of the right-hand side of eq (74) instead of p in eq (63) yields the maximal possible productivity of the system (where $\bar{\alpha}$ is chosen according to the type of contact between the transformer and reservoir). We denote

$$B = RT \sum_j \gamma_j^2 \sum_i x_{ji} \ln \frac{x_{ji}}{x_{0i}}, \quad D = T \sum_j \gamma_j^2 \sum_i \ln \frac{x_{ji}^2}{\alpha_{ji}} \qquad (76)$$

We obtain

$$p_{max} = \bar{\alpha}(\sqrt{T_+} - \sqrt{T_-})^2 = B g_{max} + D g_{0max}^2$$

and the limiting productivity is

$$g_{0max} = \frac{-B + \sqrt{B^2 + 4\bar{\alpha} D(\sqrt{T_+} - \sqrt{T_-})^2}}{2D} \qquad (77)$$

Formulas (76) and (77) allow us to estimate the limiting productivity of a heat-driven separation process for Newton's laws of heat transfer between the working body and reservoirs and mass transfer proportional to the differentials in chemical potentials (mass transfer is close to isothermal with the temperature T).

### 2.5.1. Example

Consider heat-driven monoethanamide gas cleansing. One of the components is absorbed by the cold solution from the input gas mixture. This solution is then heated and this component is vaporized. The input mixture's parameters are $\bar{T} = 350$ K, the key component's molar concentration $x = 0.5$, the rate of mixture $g_0 = 5$ mol/s. The temperatures of heat supplied/removed are correspondingly $T_h = 400$ K, $T_c = 300$K, and the heat transfer coefficients are $\alpha_+ = 8.368$ kJ/(s K) and $\alpha_- = 16.736$ kJ/(s K). The concentrations of the key components in the output flows are $x_1 = 0.9$, $x_1 = 0.1$; the mass transfer coefficients for each of the components (integral values over the whole contact surface) for the hot and cold reservoir's contacts are $\alpha_1 = 0.07$ mol²/(kg s), $\alpha_2 = 0.03$ mol²/(kg s).

Because the solution circulates and is heated and cooled in turns, the limiting power for transformation of heat into work is given by the expression (74) with the corresponding $\bar{\alpha}$

$$p_{max} = 20.711 \text{ kJ/s}$$

The power for separation is given by eq (63).

We have

$$p^0 = RT g_0 \sum_{j=1}^m \gamma_j \sum_i x_{ji} \ln \frac{x_{ji}}{x_{0i}} = 5.397 \text{ kJ/s}$$

The minimal work required for a system with Onsanger's equations are (see eq (63))

$$\Delta p = g_0^2 \sum_{j=1}^m \gamma_j^2 \sum_i \frac{x_{ji}^2}{\alpha_{ji}} = 7.238 \text{ kJ/s}$$

Thus, $p = p^0 + \Delta p = 12.636 \text{kJ/s} < p_{max}$. The work needed for separation does not exceed the maximal possible value for given heat transfer coefficients.

Let us estimate the minimal heat consumption. From eq (75) we get

$q_+ = 32.426 \text{ kJ/s}$

If the temperatures of the input and output flows are not the same then the minimal energy required for separation can be estimated using the thermodynamic balance equations (31) and (32) and the expression for $\sigma^{min}$ eq (49).

### 2.6. Conclusion

New irreversible estimates of the in-principle limiting possibilities of separation processes are derived in this paper. They take into account the unavoidable irreversibility caused by the finite rate of flows and heat and mass transfer coefficients. They also allow us to estimate the limiting productivity of a heat-driven separation and to find the most energy efficient separation sequence/regime of separation for a multicomponent mixture.

# 3. Optimization of membrane separations

## 3.1. Introduction

As the properties of membranes improve, the membrane separation of liquids and gases is more widely used in chemical engineering [8,10,11,20]. Since the mathematical modeling of membrane separations is simpler than that for most of the other separation processes, they could be controlled by varying the pressure, contact surface area, and the like during the separation process.

The minimal work needed to separate mixtures into pure components or into mixtures of given compositions can be minorized using well-known relationships of reversible thermodynamics [15]. However, this estimate is not accurate because it ignores the mass transfer laws and the properties of membranes, process productivity, possible intermediate processes of mixing, and so on. The estimates based on reversible thermodynamics are not suitable for determining the optimal sequence of operations in the separation of multicomponent systems, because they depend only on the compositions of feeds and end products and do not reflect the sequence of operations in which the end product was obtained. The work needed for separation consists of its reversible work and irreversible energy losses. The losses are equal to $\Delta S T$, where $\Delta S$ is the increment of the system entropy due to the irreversibility of the process. Below, the minimum possible production of entropy (that is, the minimal additional separation work) will be found for the separation of one component at a specified production rate and transport coefficients. Also, we will determine the dependence of this minimum on the input data for one or another process flowsheet at a fixed production rate.

## 3.2. Batch membrane separation

We will first consider a batch separation of a mixture in a system consisting of two chambers separated by a membrane permeable to only one active (to be separated) component of the mixture (Fig. 5). Let and $G_i(t)$, $C_i(t)$, $\mu_i(C_i$, and $P_i)$ denote the amount, the concentration of the active component, and its chemical potential in chamber $I$, respectively. These parameters can be varied during the process. At the initial moment of time $(t=0)$, the parameters that are specified include the mixture amount $G_1(0)=G_{10}$ in chamber 1 and the concentration $C_1(0)=C_{10}$ of the active component passing through membrane 3 at a mass transfer rate $g$, which depends on its chemical potentials on both membrane sides, $\mu_1$ and $\mu_2$. In turn, the potentials depend on the variation of the pressure and mixture composition in the first and second chambers. The pressure in the first chamber can be varied using piston 4. The process is isothermal, and the temperature $T$ is specified and remains unchanged.

The intensive variables in the second chamber are the pressure $P_2(t)$ and the chemical potential $\mu_2(t)$, which varies with time due to the accumulation of the active component in the chamber and the variation of the external conditions. Assume that the laws of this variation are known. The specification of the initial composition of the mixture $C_{10}$, the number of moles $G$ of the component that passed through the membrane in time $\tau$, and the initial number of moles $G_{10}$ determines the final composition in the first chamber,

$$C_1(\tau) = \frac{G_{10}C_{10}-G}{G_{10}-G},$$

and, hence, the reversible work of separation, which is equal to the increment of the free energy of the system:

$$A^0 = -G_{10}RT[C_{10}\ln C_{10} + (1-C_{10})\ln(1-C_{10})] + (G_{10}-G)RT\{C_1(\tau)\ln C_1(\tau) + [1-C_1(\tau)]\ln[1-C_1(\tau)]\}. \quad (78)$$

Consequently, the minimum of the produced work corresponds to the minimum of the irreversible losses of energy, which is proportional to $\Delta S$.

The increment of entropy in the system, the minimum of which should be determined for a separation process of duration $\tau$, is equal to the product of the flux and driving force:

$$\Delta S = \frac{1}{T}\int_0^\tau g(\mu_1,\mu_2)(\mu_1-\mu_2)dt \rightarrow \min. \quad (79)$$

The amount of the active component that passed through the membrane is written as

$$G = \int_0^\tau g(\mu_1,\mu_2)dt. \quad (80)$$

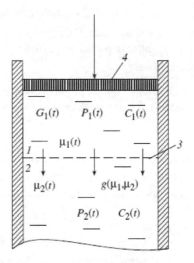

**Figure 5.** Batch separation of a mixture: *1*, chamber with the mixture to be separated; *2*, chamber to which the active component passes; *3*, membrane; *4*, piston.

The process duration $\tau$ will be fixed.

The variation of $G_1$ and concentration $C_1$ are determined by the equation:

$$\frac{d(G_1 C_1)}{dt} = \frac{dG_1}{dt} = -g(\mu_1, \mu_2). \tag{81}$$

It follows from Eq. (81) that $\dfrac{d[G_1(1-C_1)]}{dt} = 0$, implying that $G_1(t)(1-C_1(t)) = const$ for any moment of time. The latter is equal to the amount of the "inert" component of the mixture in the first chamber. It will be denoted as $\tilde{G} = G_1(0)[1 - C_1(0)]$.

The solution of Eq. (81) determines the dependence of the mixture amount in the first chamber on the active component concentration $G_1(C_1)$:

$$G_1(t) = \frac{\tilde{G}}{1 - C_1(t)}. \tag{82}$$

After expression (82) is substituted into Eq. (81), the latter takes the form

$$\frac{dC_1}{dt} = -\frac{1}{\tilde{G}_1}(1 - C_1)^2 g(\mu_1, \mu_2), \quad C_1(10) = C_{10}. \tag{83}$$

First, we will find such time variation of, chemical potential $\mu_1(t)$ that the increment of entropy takes the minimum value at a specified value of G. Then, for a specific form of chemical potential, we will find the time variation of pressure $P_1(t)$ corresponding to the found optimal variation of the chemical potential.

We will write the Lagrangian function F for the problem given by Eqs. (79) and (80) in view of the fact that the constant factor 1/T does not affect the optimality condition:

$$F = g(\mu_1, \mu_2)(\mu_1 - \mu_2 - \lambda).$$

The mass transfer rate g is equal to zero when $\mu_1 = \mu_2$ and increases monotonically with increasing $\mu_1$. As a result, the function F is, as a rule, convex with respect to $\mu_1$. Consequently, this dictates the stationary of F in the solution of the problem and this solution is unique:

$$\frac{\partial F}{\partial \mu_1} = 0 \to \frac{\partial g}{\partial \mu_1}(\mu_1 - \mu_2 - \lambda) = -g(\mu_1, \mu_2).$$

To cancel out $\lambda$, we integrate the both sides of this equality from zero to $\tau$ in view of Eq. (80) to obtain

$$\lambda = \frac{G + \int_0^\tau \frac{\partial g}{\partial \mu_1}(\mu_1 - \mu_2)dt}{\int_0^\tau \frac{\partial g}{\partial \mu_1}dt}.$$

Consequently, to determine $\mu_1(t)$ with a convex function F, we have the equation determining the optimal variation of $\mu_2(t)$ in the function $g(\mu_1, \mu_2)$ for any $\mu_1(t)$ and mass transfer law $\mu_2(t)$:

$$g(\mu_1, \mu_2) = \frac{\partial g}{\partial \mu_1}\left[ \frac{G + \int_0^\tau \frac{\partial g}{\partial \mu_1}(\mu_1 - \mu_2)dt}{\int_0^\tau \frac{\partial g}{\partial \mu_1}dt} - (\mu_1 - \mu_2) \right]. \tag{84}$$

If the flux is proportional to the difference of chemical potentials,

$$g = \alpha(\mu_1 - \mu_2), \tag{85}$$

it follows from optimality condition (84) that

Minimum Dissipation Conditions of the Mass Transfer and Optimal Separation Sequence
Selection for Multicomponent Mixtures I

93

$$\mu_1^*(t) = \mu_2(t) + \frac{G}{\tau\alpha}, \quad g^* = \frac{G}{\tau} = const. \tag{86}$$

The variation of $P_1(t)$ corresponding to $\mu_1^*(t)$ depends on the form of the chemical potential.

For mixtures close in properties to ideal gases, the chemical potential (molar Gibbs energy) of the active component of the mixture is written as

$$\mu_1 = \mu_{11}(T, P_1) + RT \ln C_1 = \mu_{01}(T) + RT \ln P_1 + RT \ln C_1, \tag{87}$$

where $\mu_{01}$ is the standard chemical potential for $P_1 = C_1 = 1$.

The variation of $C_1^*(t)$ (t) is determined by Eq. (83) with known mass transfer rate $g$. After $C_1^*(t)$ and $\mu_1^*(t)$ are substituted into Eq. (87), we obtain an expression for the pressure in the first chamber:

$$P_1^*(t) = \frac{1}{C_1^*(t)} \exp\left[\frac{\mu_1^*(t) - \mu_{01}(T)}{RT}\right]. \tag{88}$$

For the flux defined by Eq. (85) and defined by Eq. (86), Eq. (83) takes the form:

$$\frac{dC_1}{dt} = -\frac{G(1-C_1)^2}{\tau \tilde{G}_1} = \frac{G(1-C_1)^2}{G_{10}(1-C_{10})\tau}, \quad C_1(0) = C_{10}.$$

The solution to this equation is written as

$$C_1^*(t) = \frac{G_{10}C_{10} - \dfrac{G}{\tau}t}{G_{10} - \dfrac{G}{\tau}t}. \tag{89}$$

Substituting the latter into Eq. (88) gives the time variation of the pressure:

$$P_1^*(t) = \frac{G_{10} - \dfrac{G}{\tau}t}{G_{10}C_1(0) - \dfrac{G}{\tau}t} \exp\left(\frac{\mu_2 + \dfrac{G}{\alpha\tau} - \mu_{10}(T)}{RT}\right).$$

After the optimal variation of $\mu_1^*(t)$, or optimal value of this chemical potential, is found, we can determine $\Delta S_{\min}$ by substituting $\mu_1^*$ and $\mu_2$ into Eq. (79). Using the flux defined by Eq. (85) and relationship (79), we obtain

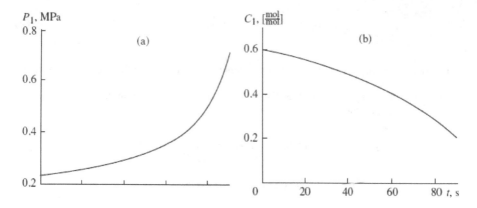

**Figure 6.** Optimal time variation of (a) pressure and (b) mole fraction of the active component in the first chamber for a gas mixture.

$$\Delta S_{min} = \frac{G^2}{T\alpha\tau}.$$

The optimal variation of the pressure and mole fraction of oxygen in the first chamber is shown in Fig. 6. It corresponds to the separation of a gas mixture composed of carbon dioxide, 120 moles of $CO_2$, and oxygen, 180 moles of $O_2$ (active component), when $G_{10}=300$ moles, $C_{10}=0.6$, $G=150$moles, $\tau=90$s, $\alpha=2.13\times10^{-3}$mol$^2$/(s J), $P_2=101330$Pa, $C_2=1.0$, and $T=283$ K. At the moment when the process is terminated, $C(\tau)=0.2$. The production of entropy is $\Delta S=\Delta S_{min}=415$ J/K.

The produced work is $A=A^0+T\Delta S=415730$ J, where according to Eq. (78) $A_0=298300$J.

Although the chemical potential for ideal solutions is written like Eq. (87), the function $\mu_{11}(T, P_1)$ for them takes a different form. This is caused by the fact that the chemical potential $\mu_1(T, P_1, x_1)$ is the molar Gibbs energy of the active component and the derivative of the chemical potential with respect to pressure is the molar volume of this component $v_1$ [15]. In contrast to gases, the molar volume of liquids is virtually independent of pressure and varies vary little with temperature. As

$$\frac{\partial\mu_1}{\partial P_1} = \frac{\partial\mu_{11}}{\partial P_1} = v,$$

we obtain

$$\mu_1(T,P_1,x_1) = \mu_{01}(T) + vP_1 + RT\ln C_1. \tag{90}$$

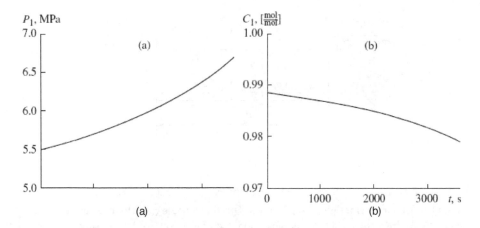

$P_1$, MPa

(a)

$C_1$, $[\frac{mol}{mol}]$

(b)

(a)

(b)

**Figure 7.** Optimal time variation of (a) pressure and (b) the mole fraction of the active component in the first chamber for a near-ideal solution.

For the flux defined by Eq. (85) and $\mu_1^*$ defined by Eq. (86), the variation of $C_1^*(t)$ for liquids can be written in the same way as for gases in Eq. (89). After $\mu_1^*(t)$ and $C_1^*(t)$ are substituted into Eq. (90), we obtain an equation for the optimal variation of pressure in the first chamber:

$$P_1^*(t) = \frac{1}{v_1}\left[\mu_2 + \frac{G}{\tau\alpha} - \mu_{01}(T) - RT\ln\left(\frac{G_{10}C_1(0) - \frac{G}{\tau}t}{G_{10} - \frac{G}{\tau}t}\right)\right].$$

For illustration, we considered the separation of water with a high salt concentration. Like ocean water, it contained 36 g/l of salt (inert component). The other process parameters were $G_{10}=552.3$ moles, $C_{10}=0.989$, $G=250$moles, $\tau=3600$s, $\alpha=9.92\times10^{-4}$mol²/(s J), $P_2=101330$Pa, $C_2=1.0$, and $T=283$ K. The time variation of the optimal pressure of the liquid and the mole fraction of water in the first chamber are illustrated in Fig. 7. At the moment when the process is terminated, $C(\tau)=0.979$. The production of entropy is $\Delta S=\Delta S_{min}=61.8$ J/K. The produced work is $A=A^0+T\Delta S=26470$ J, where according to Eq. (78) $A_0=8973$J.

### 3.3. Membrane separation process distributed along the filter

The parameters of the system can vary with length rather than with time, as in the previous system. The flow diagram of this system is shown in Fig. 8. The mixture to be separated, which is characterized by a molar flux $g_1(0)=g_{10}$ and concentration $C_1(0)=C_{10}$, is continuously supplied to the first chamber, the overall length of which is $L$. As the mixture travels over the length $l$, the active component passes across the membrane into the second chamber. The concentration of the active component in the mixture to be separated at the outlet of the first chamber is $C_1(L)=C_{1L}$. The chemical potential of this component in the second chamber,

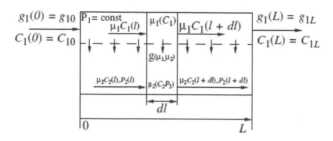

**Figure 8.** Continuous separation of a mixture.

$\mu_2(l)$, should be chosen so that in the isothermal process the increment of entropy in the system should be a minimum for the specified values of production rate $\bar{g}$ and total membrane surface area $s(L)$. In irreversible continuous separation, the power p expended for separation is the sum of the reversible component

$$p^0 = -g_1(0)RT[C_{10}\ln C_{10} + (1-C_{10})\ln(1-C_{10})] + (g_1(0)-\bar{g})RT[C_1(L)\ln C_1(L) + (1-C_1(L))\ln(1-C_1(L))], \tag{91}$$

which is determined at the given conditions, and the irreversible losses $p_H = T\sigma$. Consequently, the minimal production of entropy $\sigma$ corresponds to the minimal separation work $p$.

The flux of the component to be distributed at section $l$ is equal to $g[\mu_1(l), \mu_2(l)]$. The production rate is specified as

$$\int_0^L g(\mu_1, \mu_2)dl = \bar{g}. \tag{92}$$

The production of entropy is determined by the expression

$$\sigma = \frac{1}{T}\int_0^L g(\mu_1, \mu_2)(\mu_1 - \mu_2)dl \to \min_{\mu_2(l)}. \tag{93}$$

Assume that $\mu_2(l)$ is the control parameter.

If the operating regime in the first chamber is close to plug flow, the material balance equations for section $l$ give equations analogous to Eqs. (81).

$$\frac{d}{dl}(C_1 g_1) = \frac{d}{dl}g_1 = -g(\mu_1, \mu_2). \tag{94}$$

The above equation can be used to obtain a relationship analogous to Eq. (83):

Minimum Dissipation Conditions of the Mass Transfer and Optimal Separation Sequence
Selection for Multicomponent Mixtures I

97

$$\frac{dC_1}{dl} = -\frac{(1-C_1)^2}{\tilde{g}_1} g(\mu_1, \mu_2),$$

$$C_1(0) = C_{10}, \quad g_1(l) = \frac{\tilde{g}_1}{1 - C_1(l)},$$

$$(95)$$

where $\tilde{g}_1 = g_1(0)(1 - C_{10})$ is the molar flux of the inert component through the first chamber.

Equations (92), (93), and (95) represent an optimal control problem in which $C_1$ is the state coordinate and the potential $\mu_2$ is the control action. This problem can be simplified using the fact that for optimal processes the right-hand side of Eq. (95) never change the sign and $C_1$ monotonically varies with time. The independent variable l can be replaced by $C_1$. It follows from Eq. (95) that

$$dl = -\frac{\tilde{g}_1 dC_1}{(1-C_1)^2 g(\mu_1, \mu_2)}.$$

In view of this replacement, the problem given by Eqs. (92), (93), and (95) can be written as

$$\sigma = \frac{\tilde{g}_1}{T} \int_{C_{1L}}^{C_{10}} (\mu_1 - \mu_2) \frac{dC_1}{(1-C_1)^2} \to \min_{\mu_2}$$

$$(96)$$

with the constraints

$$\int_{C_{1L}}^{C_{10}} \frac{dC_1}{(1-C_1)^2} = \frac{\bar{g}}{\tilde{g}_1},$$

$$(97)$$

$$\int_{C_{1L}}^{C_{10}} \frac{dC_1}{(1-C_1)^2 g(\mu_1, \mu_2)} = \frac{L}{\tilde{g}_1}.$$

$$(98)$$

The concentration $C_{1L}$ is determined by the initial concentration $C_{10}$ and production rate $\bar{g}$ in constraint (97) or (94). Using constraint (94), we obtain

$$C_{1L} g_1(L) = C_{10} g_1(0) - \bar{g}, \quad g_1(L) = g_1(0) - \bar{g}$$

and, hence,

$$C_{1L} = \frac{C_{10} g_1(0) - \bar{g}}{g_1(0) - \bar{g}}.$$

$$(99)$$

The same follows from constraint (97) with $\tilde{g}_1[C_{10}, g_1(0)]$. Consequently, after $C_{1L}$ is found using constraint (99), constraint (97) can be ignored.

In distinction to batch membrane processes, the control action in a continuous membrane separation can be additionally represented by the coefficient of heat transfer $\alpha(l)$, because the membrane surface area can be varied from section to section, which corresponds to the variation of heat transfer coefficient $\alpha$. Let $\alpha$ be a function of $C_1$. After $\alpha(C_1)$ and $C_1(l)$ are found, we can pass to $\alpha(l)$. The mass transfer equation can be written as

$$g(\mu_1, \mu_2) = \alpha g_0(\mu_1, \mu_2) \tag{100}$$

where $g_0$ is called the specific mass transfer rate. In this case, the total surface area of the membrane and, hence, the overall value of the heat transfer coefficient will be bounded:

$$\int_{C_{1L}}^{C_{10}} \alpha(C_1)dC_1 = \bar{\alpha}. \tag{101}$$

In constraint (98), the mass transfer rate can be written as Eq. (100), and equality (101) can be added to the constraints of the problem. The resulting problem, given by Eqs. (96), (98), and (101), is an isoperimetric variation problem. The necessary condition for the optimality of its solution is the requirement that the Lagrangian function should be stationary with respect to $\mu_2$ and $\alpha$:

$$F = \frac{1}{(1-C_1)^2}\left[\mu_1(C_1) - \mu_2 - \frac{\lambda_1}{\alpha g_0(\mu_1, \mu_2)}\right] - \lambda_2\alpha,$$

where the multipliers $\lambda_1$ and $\lambda_2$ correspond to constraints (98) and (101). The conditions for the stationary of $F$ with respect to the desired variables are written as

$$\frac{\partial F}{\partial \mu_2} = 0 \rightarrow \frac{\lambda_1 \partial g_0/\partial \mu_2}{\alpha g_0^2(\mu_1, \mu_2)} = 1,$$

$$\frac{\partial F}{\partial \alpha} = 0 \rightarrow \frac{1}{(1-C_1)^2}\frac{\lambda_1}{\alpha^2 g_0(\mu_1, \mu_2)} = \lambda_2.$$

The above equations give the process optimality conditions:

$$\frac{\lambda_1 \partial g_0/\partial \mu_2}{\alpha(C_1)g_0^2(\mu_1, \mu_2)} = const = \frac{1}{\lambda_1}, \tag{102}$$

Minimum Dissipation Conditions of the Mass Transfer and Optimal Separation Sequence
Selection for Multicomponent Mixtures I

99

$$\alpha^2(C_1)(1-C_1)^2 g_0(\mu_1,\mu_2) = const = \frac{\lambda_1}{\lambda_2}. \tag{103}$$

From constraints (98) and (103) we obtain

$$\frac{\lambda_1}{\lambda_2} = \frac{\tilde{g}_1 \int\limits_{C_{1L}}^{C_{10}} \alpha(C_1)dC_1}{L} = \frac{\tilde{g}_1 \bar{\alpha}}{L}. \tag{104}$$

It follows from (101) and (102) that

$$\frac{1}{\lambda_1} = \frac{1}{\bar{\alpha}} \int\limits_{C_{1L}}^{C_{10}} \frac{\partial g_0/\partial \mu_2}{g_0^2(\mu_1,\mu_2)} dC_1. \tag{105}$$

After expressions (105) and (104) are substituted into conditions (102) and (103), respectively, we can use the known function $\mu_1(C_1)$ to find the functions $\mu_2^*(C_1)$ and $\alpha^*(C_1)$ that are optimal in terms of minimal irreversibility, which with the help of Eq. (95) determine $C_1^*(l)$ and, hence, $\mu_2^*(l)$ and $\alpha^*(l)$.

Let us write the above relationships specifically for the function g written as a linear function of the difference of chemical potentials, Eq. (85), and chosen functions $\mu_i(C_i)$. Assume that the specific mass transfer rate takes the form:

$g_0(\mu_1,\mu_2) = \mu_1 - \mu_2.$

Constraints (102)–(105) lead to the equations

$$\frac{1}{\alpha(C_1)(\mu_1(C_1) - \mu_2(C_1))^2} = \frac{1}{\bar{\alpha}} \int\limits_{C_{1L}}^{C_{10}} \frac{dC_1}{(\mu_1(C_1) - \mu_2(C_1))^2}, \tag{106}$$

$$\alpha(C_1)(1-C_1)^2 [\mu_1(C_1) - \mu_2(C_1)] = \frac{\tilde{g}_1 \bar{\alpha}}{L}. \tag{107}$$

For brevity, we will introduce the notation $\mu_1 - \mu_2 = \Delta\mu$ and the right-hand sides in constraints (106) and (107) will be denoted as $R_1$ and $R_2$. In this case, the above equations can be written as

$$\frac{1}{\alpha \Delta \mu^2} = R_1, \quad \alpha^2 (1 - C_1) \Delta \mu = R_2,$$

and we obtain

$$\Delta \mu^*(C_1) = \frac{(1 - C_1)^{2/3}}{R_2^{1/3} R_1^{2/3}}, \quad \alpha^*(C_1) = \frac{R_1^{1/3} R_2^{2/3}}{(1 - C_1)^{4/3}}. \tag{108}$$

The concentration of the active component in the first chamber declines with increasing l. Therefore, under optimal operating conditions, $\Delta \mu^*(l) = \Delta \mu^*[C_1(l)]$ increases while the surface area of the membrane, which is proportional to $\alpha^*(l) = \alpha^*[C_1(l)]$, decreases.

To find $R_1$, we will substitute Eq. (108) into Eq. (106) to obtain the equation

$$\frac{R_1^{4/3} R_2^{2/3}}{\bar{\alpha}} \int_{C_{1L}}^{C_{10}} \frac{dC_1}{(1 - C_1)^{4/3}} = R_1.$$

The evaluation of the integral gives us the desired formula:

$$R_1 = \frac{\bar{\alpha} L^2}{B^3 \tilde{g}_1^2}, \tag{109}$$

where

$$B = \left( \frac{3}{\sqrt[3]{1 - C_1(0)}} - \frac{3}{\sqrt[3]{1 - C_1(L)}} \right).$$

Equation (108) in view of Eq. (109) yields the optimal dependence of the difference of chemical potentials on the concentration $C_1$:

$$\Delta \mu^*(C_1) = \frac{(1 - C_1)^{2/3} B^2 \tilde{g}_1}{\bar{\alpha} L}, \tag{110}$$

$$\alpha^*(C_1) = \frac{\bar{\alpha}}{B(\sqrt[3]{1 - C_1})^4}. \tag{111}$$

Consequently, Eq. (95) takes the form:

Minimum Dissipation Conditions of the Mass Transfer and Optimal Separation Sequence
Selection for Multicomponent Mixtures I

101

$$\frac{dC_1}{dl} = -\frac{(1-C_1)^2}{\tilde{g}_1} a^*(C_1) \Delta \mu^*(C_1) = -\frac{(1-C_1)^{4/3}B}{L}.$$

Integrating this equation with specified initial conditions, we can find the variation of the concentration of the active component over the length of the first chamber under optimal operating conditions:

$$C_1^*(l) = 1 - \frac{27L^3}{\left(\dfrac{3L}{\sqrt[3]{1-C_1(0)}} - Bl\right)^3}. \tag{112}$$

Substituting this expression into Eqs. (110) and (111) yields the variation of the desired variables over the length:

$$\Delta \mu^*(l) = \frac{9LB^2 \tilde{g}_1}{\bar{\alpha}\left(\dfrac{3L}{\sqrt[3]{1-C_1(0)}} - Bl\right)^2}, \tag{113}$$

$$\alpha^*(l) = \frac{\bar{\alpha}\left(\dfrac{3L}{\sqrt[3]{1-C_1(0)}} - Bl\right)^4}{3^4 BL^4}. \tag{114}$$

The minimal value of the production of entropy corresponding to the above solution is written as

$$\sigma_{min} = \frac{\tilde{g}}{T} \int_{C_{1L}}^{C_{10}} \Delta \mu^*(C_1) \frac{dC_1}{(1-C_1)^2} = \frac{\tilde{g}_1^2}{LT\bar{\alpha}} B^3. \tag{115}$$

We will introduce $\alpha$, the mass transfer coefficient per unit area of the membrane surface, and ds(l), the elementary membrane surface area. If $\alpha = constant$, then

$$\bar{\alpha} = \alpha \int_0^L ds(l) = \alpha s(L),$$

where $s(L)$ is the total contact surface area.

If the specific mass transfer coefficient of the membrane material $\alpha$ and the total contact surface area $s(L)$ are known, we can find the optimal distribution of the membrane surface area over the length of the filter:

$$ds^*(l) = \frac{\alpha^*(l)}{\bar{\alpha}}.$$

For near-ideal gas mixtures, we can write

$$\mu_1(C_1) = \mu_{10}(P_1, T) + RT \ln C_1,$$

where $P_1$ and $T$ are assumed to be specified, and

$$\mu_2(P_2, T, C_2) = \mu_{20}(T) + RT \ln P_2 + RT \ln C_2. \tag{116}$$

When $\Delta\mu^*(C_1)$ is known, expression (116) can be used to find the pressure function in the second chamber for which $\sigma_{min}$ is achieved:

$$P_2^*(C_1, C_2) = \frac{1}{C_2} \exp\left\{ \frac{1}{RT}[\mu_1(C_1) - \mu_{20}(R) - \Delta\mu^*(C_1)] \right\}. \tag{117}$$

The optimal curves for the pressure and mass transfer coefficient are plotted in Fig. 9, in which the data refer to the separation of a gas mixture composed of carbon dioxide $CO_2$ and oxygen $O_2$ (active component) when $C_{10} = 0.6$, $g_1(0) = 3.33$mol/s, $\bar{g} = 1.66$mol/s, $P_1 = 2 \times 10^6$Pa, $\bar{a} = 3.18 \times 10^{-4}$ mol²/(s J), $\alpha = 0.013$mol²/(s J), $L = 2.5$m, and $T = 283$ K.

At the filter outlet, $C_1(L) = 0.2$. The production of entropy is $\sigma = \sigma_{min} = 4.6$ J/(s K).

The consumed power is $p = p_0 + T\sigma = 4600$ J/s, where according to Eq. (91) $p_0 = 3292$J/s.

For ideal solutions, the calculation is almost the same except for the form in which the chemical potentials are written. For the first chamber,

$$\mu_1(C_1) = \mu_{10}(T) = vP_1 + RT \ln C_1,$$

where $v$ is the molar volume of the active component.

For the second chamber,

$$\mu_2(P_2, T, C_2) = \mu_{20}(T) + vP_2 + RT \ln C_2.$$

The dependence of the solution pressure in the second chamber on the concentration is written as

$$P_2^*(C_1, C_2) = \frac{1}{v}[\mu_1(C_1) - \mu_{20}(T) - \Delta\mu^*(C_1) - RT \ln C_2].$$

Minimum Dissipation Conditions of the Mass Transfer and Optimal Separation Sequence
Selection for Multicomponent Mixtures I

103

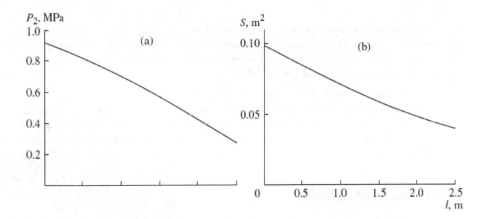

**Figure 9.** Optimal variation of (a) pressure and (b) the membrane surface area over the filter length for a near-ideal gas mixture.

For illustration, we considered the separation of water with a high salt concentration. Like ocean water, it contained 36 g/l of salt (inert component). The other process parameters were $g_1(0)=0.153$ mol/s, $C_{10}=0.989$, $\bar{g}=1.66$ mol/s, $\bar{a}=2.92\times10^{-6}$ mol²/(s J), $\alpha=0.011$ mol²/(s J), $P_1=7.7\times10^6$ Pa, $L=3$m, and $T=283$ K. The profile of optimal pressure in the first chamber and the variation of the mass transfer coefficient over the filter length are illustrated in Fig. 10. At the filter outlet, $C_1(L)=0.979$. The production of entropy is $\sigma=\sigma_{min}=0.017$ J/(s K).

The consumed power is $p=p_0+T\sigma=7.35$ J/s, where according to Eq. (91) $p_0=2.47$ J/s.

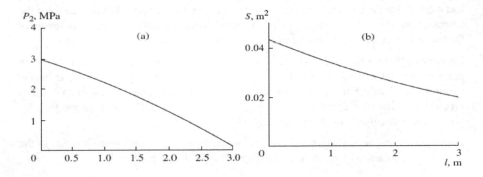

**Figure 10.** Optimal variation of (a) pressure and (b) membrane surface area over the filter length for a near-ideal solution.

### 3.4. Conclusion

The minimal losses of energy for irreversible membrane separations with specified production rates are estimated. The variation of the driving force (difference of chemical potentials) and the distribution of the membrane surface area over the filter length corresponding to the process with minimal energy losses are found.

The obtained estimates can be used for assessing the deviation of the actual membrane separation from the optimal process and for comparing the thermodynamic efficiency of membrane separation processes with different flow diagrams, as well as for formulating and solving problems regarding the optimal sequence of operations in the separation of multicomponent mixtures.

## 4. Optimization of diffusion systems

### 4.1. Introduction

The problem of deriving work from a irreversible thermodynamic system and the inverse problem of maintaining its irreversible state by consuming energy are central in thermodynamics. For systems that are not in equilibrium with respect to temperature, the first (direct) of the above problems is solved using heat engines and the second one (inverse) is solved using heat pumps. For systems that are not in equilibrium with respect to composition, the second problem is solved using separation systems and the first one is solved using diffusion engines. As a rule, separation systems and diffusion engines are based on membranes.

There is a lot of studies of membrane separation systems and diffusion engines in the literature [5,7]. In the present paper, these systems will be considered using the theory of finite-time thermodynamics. The finite-time thermodynamics, which evolved in the past years, studies the limiting performance of irreversible thermodynamic systems when the duration of the processes is finite and the average rate of the streams is specified [14, 17]. For example, some problems for heat engines, such as maximizing the power at given heat transfer coefficients and maximizing the efficiency at given power for different conditions of contact between the working body and surroundings, are already solved. In this case, the irreversible processes of the interaction of subsystems each of which is in internal equilibrium are considered.

For systems that are not uniform in concentration, it is most important to study the limiting performance of separation systems. In this case, however, the inverse problem of studying the performance of diffusion engines is of definite interest as well. The simplest variant of this problem was first formulated by Rozonoer [17]. The review of the literature shows that this problem was discussed rather superficially.

In the present paper, we will study the limiting performance of membrane systems in the separation processes with fixed rates, focusing on the following problems:

1.  Minimizing the amount of energy necessary for the separation of a feed mixture with a given composition into separation products with given compositions at a given average production rate.

**2.**   Maximizing the power and efficiency of diffusion engines.

The solution of these problems depends strongly on whether the feed mixture used by the engine is gaseous or liquid because this determines the form of the chemical potentials of components and, hence, the driving forces of the process. For near-ideal gas mixtures, the chemical potential of component *I* of the mixture takes the form [15]:

$$\mu_i(T, P_i) = \mu_0(T) + RT \ln P_i, \quad i = 1, 2, \ldots,$$

where $P_i$ is the partial pressure of component *I* and $\mu_0(T)$ is the chemical potential of the pure component. Assuming that the ratio of the partial pressure to the total pressure is equal to $x_i$,

$$P_i = P x_i = P \frac{N_i}{N}, \quad i = 1, 2, \ldots,$$

we can rewrite the expression for the chemical potential in the form:

$$\mu_i(T, P, x_i) = \mu_1(T, P) + RT \ln x_i, \tag{118}$$

where $\mu_1(T, P) = \mu_0(T) + RT \ln P$.

Although the chemical potential for liquids has the same form as Eq. (118), the form of the function $\mu_1(T, P)$ is different. This is caused by the fact that the chemical potential $\mu_i(T, P, x_i)$ represents the molar Gibbs energy of component I and its derivative with respect to pressure is equal to the molar volume of this component $v_i$ [15]. In contrast to gases, the molar volume of liquids is virtually independent of pressure and weakly dependent on temperature. As

$$\frac{\partial \mu_i}{\partial P} = \frac{\partial \mu_1}{\partial P} = v_i,$$

we obtain

$$\mu_i(T, P, x_i) = \mu_1(T) + RT \ln x_i. \tag{119}$$

It is assumed that the processes are isothermal and the temperatures of all subsystems are equal to T. The problems listed above will be considered for gaseous mixtures and then for liquid solutions.

## 4.2. Limiting performance of diffusion systems for gaseous mixtures

### 4.2.1. Maximum work in a membrane process

Consider a system consisting of a thermodynamic reservoir, the intensive variables of which are fixed and are independent of mass transfer fluxes, and a working body, the intensive variables of which can be varied with time by one or another way. The system can consume external energy or generate work. In the first case, the work will be negative; in the second, positive.

The reservoir and the working body interact through a membrane that is permeable only to one (active) component of the mixture. The mass transfer rate $g$ depends on the chemical potentials of the active component in the reservoir $\mu_0$ and in the working body $\mu(t)$. When these chemical potentials are equal to each other, the flux is equal to zero. In the particular case under consideration,

$$g(\mu_0, \mu) = \alpha(\mu_0 - \mu), \tag{120}$$

where $\alpha$ is the mass transfer coefficient. The working-body temperature T is maintained constant and equal to the reservoir temperature.

When the process duration $\tau$ and the total amount of the component $G_0$ transferred from the reservoir to the working body and in the reverse direction are fixed in the process characterized by a finite mass transfer coefficient, the chemical potentials $\mu_0$ and $\mu(t)$ should differ from each other at every moment of time and the mass transfer process should be irreversible. For definiteness, we assume that $\mu_0 > \mu(0)$ and that the component is transferred from the reservoir to the working body.

The variation of the system entropy will be caused by the decrease in the reservoir entropy, the increase in the entropy of the working body, and the production of entropy due to the irreversible mass transfer $\sigma$. For a given initial state of the system (that is, the compositions of mixtures at the initial moment of time, the total amount of the substance in the working body) and a given constant value of the quantity

$$G_0 = \int_0^\tau g[\mu_0, \mu(t)]dt \tag{121}$$

the variation of the entropies of the reservoir and working body with time $\tau$ are completely determined and the minimal increase in the system entropy corresponds to the minimum of the entropy production:

$$\bar{\sigma} = \frac{1}{T}\int_0^\tau [\mu_0 - \mu(t)]g[\mu_0, \mu(t)]dt \to \min. \tag{122}$$

In this case, the function $\mu(t)$ should be chosen.

Let us find the quantitative relationship between the work $A$, which can be extracted (consumed) in this process, and the value of $\bar{\sigma}$. For simplicity, we assume that the mixture in the reservoir and working body consists of two components (a more general case can similarly be considered by introducing an equivalent component). If the concentrations of the active component in the reservoir and working body are $x_0$ and $x(t)$, the concentrations of the second component will be equal to $1-x_0$ and $1-x(t)$, respectively. The variation of the substance amount $G$ and the concentration $x(t)$ of the active component in the working body are determined by the differential equations:

$$\frac{d}{dt}(Gx) = \frac{dG}{dt} = g(\mu_0, \mu), \quad G(0) \to fix,$$
$$x(0) \to fix. \tag{123}$$

As the amount of the second component is maintained constant, we obtain

$$G(0)[1 - x(0)] = [G(0) + G_0][1 - x(\tau)]. \tag{124}$$

It follows from (123) and (124) that

$$\frac{dx}{dt} = \frac{1}{G(0)[1 - x(0)]}(1 - x)^2 g(\mu_0, \mu),$$
$$x(0) \to fix. \tag{125}$$

The equations for the material, energy, and entropy balances around the system take the form:

$$G_0 x_0 = G(\tau)x(\tau) - G(0)x(0), \tag{126}$$

$$G_0 h_0 - [G(\tau)h(\tau) - G(0)h(0)] = A, \tag{127}$$

$$G_0 s_0 = [G(\tau)s(\tau) - G(0)s(0)] + \bar{\sigma} = 0, \tag{128}$$

where $h_0$ and $h$, $s_0$ and $s$ are the molar enthalpies and entropies of the mixture in the working body and reservoir, respectively. They are related by the equation [15]:

$$s = \frac{1}{T}\left(h - \sum_{i=1}^{2} \mu_i x_i\right), \tag{129}$$

$$s_0 = \frac{1}{T}\left(h_0 - \sum_{i=1}^{2} \mu_{i0} x_{i0}\right). \tag{130}$$

The pressure in the working body can vary with time, provided that $P(0)=P(\tau)$. For the chemical potentials defined by Eq. (118), the equation of entropy balance (128) in view of (127), (129), and (130) can be rewritten as

$$
\begin{aligned}
A/T = &-\bar{\sigma} + R\{G_0[x_0 \ln x_0 + (1-x_0)\ln(1-x_0)] \\
&+G(0)[x(0)\ln x(0) + (1-x(0))\ln(1-x(0))] \\
&-G(\tau)[x(\tau)\ln x(\tau) + (1-x(\tau))\ln(1-x(\tau))]\}.
\end{aligned}
\tag{131}
$$

The second term in the right-hand side of this equality can be calculated using $G_0$, $x_0$, $G(0)$, and $x(0)$. The latter ones are related through (124) and (126) to the values of $G(\tau)$ and $x(\tau)$. Let us denote the second term as $B[G_0, x_0, G(0), x(0)]$. It can be either positive or negative. It follows from equality (131) that

$$
A_{max} = T(B - \bar{\sigma}_{min}).
\tag{132}
$$

The maximum of the produced (minimum of the spent) work corresponds to the minimum of entropy production in the mass transfer process.

The problem of finding the minimum of $\bar{\sigma}$ when constraint (121) is valid (or the equivalent problem for the maximum of $G_0$ at a given constant value of $\bar{\sigma}$) is an averaged nonlinear programming problem [22]. Unlike the problem for the constrained maximum of a function, its optimal solution can vary with time. This solution is a piecewise constant function that can take not more than two values. We will not calculate these values and the fraction of the whole process time during which $\mu^*(t)$ takes each of these values because in the most common case, where the Lagrangian function for the unaveraged problem

$$
L = g[\mu_0, \mu(t)][\mu_0 - \mu(t) - \lambda]
$$

is convex with respect to $\mu$ (second derivative of $L$ with respect to $\mu$ is positive), the solution to the formulated problem is constant. Consequently, the constancy condition depends on the validity of the inequality:

$$
\frac{\partial^2 g}{\partial \mu^2}(\mu_0 - \mu(t) - \lambda) - 2\frac{\partial g}{\partial \mu} \geq 0.
\tag{133}
$$

The multiplier $\lambda$, which is equal to the derivative of the minimum value of $\bar{\sigma}$ with respect to $G_0$, should be positive due to the physical nature of the problem. The second derivative of $L$ with respect to $\mu$ for the mass transfer rate in the form of (120) is equal to $2\lambda\alpha$ and is known to be positive. In all cases where inequality (133) holds, the optimal value of the chemical potential of the active component for the working body is constant and determined by the equation:

Minimum Dissipation Conditions of the Mass Transfer and Optimal Separation Sequence
Selection for Multicomponent Mixtures I

109

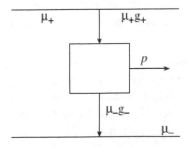

**Figure 11.** Schematic diagram of a diffusion engine with a constant contact between the working body and sources.

$$g(\mu_0, \mu^*) = \frac{G_0}{\tau}.$$ (134)

Consequently, the chemical potential of the active component of the working body for any rate satisfying (133) should be controlled so that the mass transfer rate should be constant.

The law of variation of the control variable, such as the working-body pressure, corresponding to this solution will not be constant in time because the mixture composition is varied during the process according to Eq. (125), in which the flux is determined by Eq. (134).

For mass transfer law (120), the minimal entropy produced is $\bar{\sigma}_{min} = G_0^2/\alpha\tau$. It follows from equality (132) that positive work can be extracted from the system under study only when $\tau > \tau_{min} = G_0^2/\alpha B$. It is easy to see that the process duration $\tau^*$, for which the average extraction rate of work $A^*(\tau)/\tau$ is maximal, is twice larger than $\tau_{min}$.

In the case where the system contains a source of a finite capacity at constant temperature and pressure instead of the reservoir (source of an infinite capacity), the fraction of the active component varies according to an equation similar to (125). As a result, the chemical potential $\mu_0$ is changed. However, here also, the minimum of the entropy production for mass transfer law (120) corresponds to such variation of $\mu(t)$ that the mass transfer rate is maintained constant.

Instead of the calendar time, the problem can be studied using the time of contact, when the working body moves and its parameters at every point of the loop remain constant. This can be used to determine the optimal laws of pressure variation for the zones of contact between the working body and source.

*4.2.2. Diffusion-mechanical cycle for maximum power*

Let us consider the direct cycle of work extraction in a system consisting of a working body and two reservoirs with different chemical potentials. In the first reservoir, the chemical potential of the key element is equal to $\mu_+$; in the second, $\mu_-$; for definiteness, $\mu_+ > \mu_-$ (Fig. 11).

The process is cyclic: the increase in entropy, internal energy, and mass of the key component of the working body around the cycle is equal to zero. The temperatures are the same for all subsystems.

**Alternating contact with reservoirs.** Consider the case where the working body alternately contacts the first and second reservoirs and its parameters are cyclically varied with time. Let $\tau$ stand for the cycle duration and $\mu_0(t)$ stand for the source chemical potential, which can take the values of $\mu_+$ and $\mu_-$. The formulation of the problem dealing with the production of maximum work A in a given time $\tau$ takes the form:

$$A = \int_0^\tau \mu g(\mu_0, \mu)dt \to \max_{\mu_0, \mu} \qquad (135)$$

with the constraints placed on the increment in the amount of the working-body:

$$\Delta G = \int_0^\tau g(\mu_0, \mu)dt = 0. \qquad (136)$$

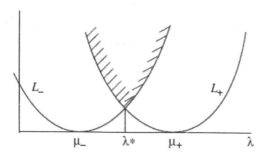

**Figure 12.** Dependence of the maximum of the Lagrangian function with respect to $\mu$ on $\lambda$.

To calculate the basic values of $\mu$ and $\mu_0$ in the problem given by (135) and (136), we can write the Lagrangian function and find its maximum with respect to $\mu_0$ and $\mu$ and its minimum with respect to $\lambda$:

$$L = \left\{ g(\mu_0, \mu)(\mu - \lambda) \right\} \to \max_{\mu_0, \mu} \min_{\lambda}.$$

The number of basic values of $\mu_0$ is equal to two: one of them corresponds to $\mu_0 = \mu_+$ and the other to $\mu_0 = \mu_-$. For the Lagrangian function $L$ that is strictly convex with respect to $\mu$, the basic values of $\mu$ satisfy the conditions:

Minimum Dissipation Conditions of the Mass Transfer and Optimal Separation Sequence
Selection for Multicomponent Mixtures I

111

$$\frac{\partial L}{\partial \mu} = \frac{\partial g}{\partial \mu}(\mu - \lambda) + g(\mu_0, \mu) = 0$$

or

$$\frac{g(\mu_0, \mu)}{(\mu - \lambda)} = -\frac{\partial g(\mu_0, \mu)}{\partial \mu}.$$

The roots for this equation for $\mu_0 = \mu_+$ and $\mu_0 = \mu_-$ will be denoted by $\mu_1$ and $\mu_2$, respectively. As $L$ is maximal at the basic points, we can write

$$L(\mu_+, \mu_1, \lambda) = L(\mu_-, \mu_1, \lambda), \tag{137}$$

which determines the value of $\lambda$.

Let us specify the obtained relations for

$g(\mu_0, \mu) = \alpha(\mu_0)(\mu_0 - \mu).$

It follows from (137) that

$$\mu_1 = \frac{\mu_+ + \lambda}{2}, \quad \mu_2 = \frac{\mu_- + \lambda}{2}. \tag{138}$$

Substituting $\mu_1$ and $\mu_2$ into the function L for each basic value gives its dependence on $\lambda$:

$$L_+ = L\ (\mu_+, \mu_1) = \frac{\alpha_+}{4}(^{\mu_+ \lambda 2}_{\ -}),$$

$$L_- = L\ (\mu_-, \mu_2) = \frac{\alpha_-}{4}(^{\mu_- \lambda 2}_{\ -}).$$

The maximum of L with respect to $\mu_0$ and $\mu$ reaches its minimal value with respect to $\lambda$ (Fig. 12) when

$$L_+(\lambda) = L_-(\lambda) \rightarrow \lambda^* = \frac{\sqrt{\alpha_+}\mu_+ + \sqrt{\alpha_-}\mu_-}{\sqrt{\alpha_+} + \sqrt{\alpha_-}}. \tag{139}$$

The fractions of time $\tau$ of contact with reservoirs are determined by Eq. (136) and can be written as

$$\gamma_+ = \frac{\alpha_-\sqrt{\alpha_+}}{\alpha_-\sqrt{\alpha_+} + \alpha_+\sqrt{\alpha_-}},$$

$$\gamma_- = \frac{\alpha_+\sqrt{\alpha_-}}{\alpha_-\sqrt{\alpha_+} + \alpha_+\sqrt{\alpha_-}}.$$

The maximal work in time $\tau$ takes the form:

$$A^*(\tau) = \tau[\gamma_+ \mu_1 \alpha_+ (\mu_+ - \mu_1) + \gamma_- \mu_2 \alpha_- (\mu_2 - \mu_-)],$$

where $\mu_1$ and $\mu_2$ can be determined from (138) after the value of $\lambda$ from (139) is substituted into this expression. The maximal power is equal to

$$\frac{A^*(\tau)}{\tau} = [\gamma_+ \mu_1 \alpha_+ (\mu_+ - \mu_1) + \gamma_- \mu_2 \alpha_- (\mu_2 - \mu_-)].$$

**Constant contact with sources**. In heat engines, there can be either alternate or constant contact between the working body and sources. In the latter case, the parameters of the working body are distributed and the process in it can be regarded close to reversible if the distribution of the parameters is caused by the conductive flux. Likewise, a constant contact with sources is possible in systems that are not homogeneous in concentration, such as separation systems and diffusion engines.

In this case, the maximal power takes the form of a nonlinear programming problem:

$$p = [g_1(\mu_+, \mu_1)\mu_1 - g_2(\mu_2, \mu_-)\mu_2] \rightarrow \max_{\mu_1, \mu_2}$$

with the constraint

$$g_1(\mu_+, \mu_1) - g_2(\mu_2, \mu_-) = 0. \tag{140}$$

The optimality constraint for this problem leads to the relation:

$$\mu_1 - \mu_2 = \frac{g_2(\mu_2, \mu_-)}{\partial g_2 / \partial \mu_2} - \frac{g_1(\mu_+, \mu_1)}{\partial g_1 / \partial \mu_1}, \tag{141}$$

which together with equality (140) determines the desired variables.
Let $g_1$ and $g_2$ are proportional to the difference between the chemical potentials:

$$g_1 = \alpha_1(\mu_+ - \mu_1), \quad g_2 = \alpha_2(\mu_2 - \mu_-).$$

Equality (141) can be written in the form:

$$\mu_1 - \mu_2 = \frac{\mu_+ - \mu_-}{2}. \tag{142}$$

The constraint $g_1 = g_2$ results in

$$\alpha_1 \mu_1 + \alpha_2 \mu_2 = \alpha_1 \mu_+ + \alpha_2 \mu_-. \tag{143}$$

The solution to Eqs. (142) and (143) can be written as

$$\mu_2^* = \frac{1}{2(\alpha_1 + \alpha_2)}\left[\mu_+\alpha_1 + \mu_-(\alpha_1 + 2\alpha_2)\right],$$

$$\mu_1^* = \frac{1}{2(\alpha_1 + \alpha_2)}\left[\mu_+(\alpha_2 + 2\alpha_1) + \mu_-\alpha_2\right].$$

The value of maximal power corresponding to this choice is

$$p_{max} = \frac{\bar{\alpha}}{4}(\mu_+ - \mu_-)^2,$$

where the equivalent mass transfer coefficient is defined as

$$\bar{\alpha} = \frac{\alpha_1\alpha_2}{\alpha_1 + \alpha_2}.$$

### 4.3. Limiting performance of diffusion systems for liquid mixtures

The result obtained above for the membrane systems consisting of a working body and a source of finite or infinite capacity using gaseous mixtures can be translated in the same form to liquid solutions with allowance for the different form of the chemical potential. Diffusion engines are most often designed for the treatment of saline water. Let us consider two flow-sheets of liquid diffusion engines.

#### 4.3.1. Diffusion engine with a constant contact between the working body and the sources

Let the system consist of two liquids with the same temperature separated by a semipermeable membrane. One of the liquids is a pure solvent and the other is a solution in which some substance of concentration $C$ is dissolved. The membrane is permeable only to the solvent. The equilibrium in the system is reached as soon as the chemical potentials calculated by formula (119) become equal to each other:

$$v_0 P_0 - v_r P_r = -RT\ln x_r.$$

Let the difference of pressure across the membrane be denoted as $\pi$. Also, we will keep in mind that the molar volumes $v_0$ and $v_r$ for low concentrations are equal to each other. The mole fraction of the dissolved component will be denoted as $x_1$. If its value is low, then $\ln x_r = \ln(1 - x_1) \approx -x_1$. In this case,

$$\pi = RT\frac{x_1}{v_0} = -RTC. \tag{144}$$

Equation (144) is called the Van't Hoff equation for osmotic pressure.

Consider the system shown in Fig. 13. The chamber to the left of the membrane contains a pure solvent at an environmental pressure equal to $P_0$. The chamber of volume V to the right of the membrane contains a continuously replenished solution in which the concentration of the

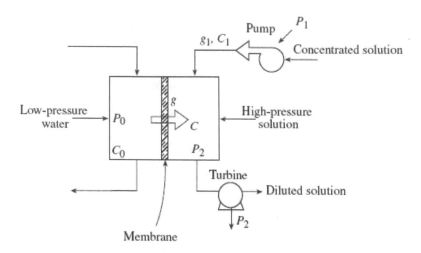

**Figure 13.** Schematic diagram of a diffusion engine with a constant contact between the working body and sources.

dissolved component is C. The pressure in the right chamber is $P_2$ and the solution is assumed to be ideal. When an equilibrium is reached in the right chamber (that is, the flux through it is equal to zero), the pressure established in it will exceed $P_0$ by the value of osmotic pressure $\pi$. The osmotic pressure value is related to the concentration and temperature in the chamber by the Van't Hoff equation. When the solution in the chamber is replenished, the pressure $P_2 < P_0 + \pi$, giving rise to a solvent flux g across the semipermeable membrane. Conventionally, the diffusion flux is taken to be equal to the difference between the actual and equilibrium pressures:

$$g = \alpha(P_0 + \pi - P_2) = \alpha(\pi - \Delta P), \tag{145}$$

where $\Delta P = P_2 - P_0$

Let $p_1$ stand for the power of the pump supplying the concentrated solution, $g_1$ stand for the flow rate of this solution, and $C_1$ stand for the solution concentration. Assuming that the pump efficiency is 100%, we obtain

$p_1 = \Delta P g_1$.

The additional flux across the membrane increases the volume of the solution, which drives a turbine and generates power $p_2$:

$p_2 = (g_1 + g)\Delta P$.

Consequently, the power r and efficiency $\eta$ of the saline diffusion engine can be written as

Minimum Dissipation Conditions of the Mass Transfer and Optimal Separation Sequence
Selection for Multicomponent Mixtures I

115

$$p = p_2 - p_1 = g\Delta P = \alpha(\pi - \Delta P)\Delta P,$$

$$\eta = \frac{P}{g_1} = \frac{\alpha(\pi - \Delta P)\Delta P}{g_1}.$$

where the diffusion engine efficiency is the work extracted from 1 m³ of the concentrated solution. From here on, according to the accepted system of units, the units of power and efficiency referred to a unit membrane surface area are $J/(m^2\ s)$ and $J/m^3$, respectively. If the relationship between $\pi$ and $\Delta P$ is ignored, the power reaches a maximum when $\Delta P = \pi / 2$ and its upper limit is written as

$$\bar{p} = \alpha\pi^2/4 = \alpha/4(CRT)^2.$$

As $C < C_1$, the value of the power is always less than

$$\bar{p}^* = \alpha / 4(C_1 RT)^2. \tag{146}$$

which is the upper bound for the maximal power.

The estimate produced by Eq. (146) can be refined if we take into consideration that $g$, $\Delta P$, and $C$ are related to each other by Eq. (145) and the equation of material balance on the dissolved component

$$(g_1 + g)C = g_1 C_1. \tag{147}$$

Expressing $C$ and $\Delta P$ in terms of g from Eqs. (145) and (147) and substituting them into $p$ and $\eta$, we obtain

$$C = \frac{g_1 C_1}{g_1 + g}, \quad \Delta P = CRT - \frac{g}{\alpha}, \tag{148}$$

$$p = g\Delta P = \frac{RTC_1 g_1 g}{g_1 + g} - \frac{g^2}{\alpha}, \tag{149}$$

$$\eta = \frac{\alpha(\pi - \Delta P)\Delta P}{g_1} = \frac{RTC_1 g}{g_1 + g} - \frac{g^2}{\alpha g_1}. \tag{150}$$

The points of maximum with respect to g for two concave functions (149) and (150) coincide. Consequently, to find the optimal value of $g^*$, we will use one of the functions, specifically the expression for p. The condition for the maximum with respect to g leads to the inequality:

$$g(g_1 + g)^2 = \frac{\alpha RT g_1^2 C_1}{2}. \tag{151}$$

Equation (151) can be rewritten as

$$\frac{g^3}{g_1^2} + 2\frac{g^2}{g_1} + g = \frac{\alpha RT C_1}{2} \tag{152}$$

and its right-hand side can be denoted for brevity as M. Its solution will be denoted as $g^*$. It is obvious that it satisfies the inequality:

$0 < g < M$.

Numerical solution of Eq. (152) makes it possible to refine the value of the limiting power of the diffusion engine and find the corresponding operating conditions. Equation (151) determines $g^*$ for the chosen values of $g_1$ and $C_1$; Eq. (148), for $C^*$ and $\Delta P^*$.

It should be noted that the ideal solution bounds the value of the concentration of the working solution:

$$C = C_1 \frac{g_1}{g_1 + g}.$$

The concentration should not be very high: otherwise, the molecules of the dissolved component will interact with each other and relation (144) is upset.

**Diffusion Engine with an Alternate Contact between the Working Body and Sources.** In the schematic diagram of the diffusion engine discussed in the preceding section, the working body was an open system working in constant contact with two sources under steady-state conditions. One of them supplied a concentrated solution and the other supplied a solvent.

Figure 14 shows the schematic diagram for a diffusion engine in which the working body alternately contacts each of the sources, receiving a solvent through one membrane and giving it up to a concentrated solution through another membrane. In this case, the pressure and flow rate of the working body are periodically varied: pressure increases for a lower flow rate (power $p_1$ is consumed) and decreases for a higher flow rate (power $p_2$ is generated).

We will write the balance equations for this diagram and study its limiting performance, ignoring the energy losses for driving the flow of the concentrated solution through the bottom chamber and assuming that the concentration of the dissolved component in the $g_2$ flow is equal to unity and that the pressure of the surrounding medium is equal to $P_0$. For simplicity, flow rates will be used instead of mole fluxes

The engine power is

$p = p_2 - p_1 = (g_1 + g)\Delta P_{21} - g_1\Delta P_{21} = g\Delta P_{21},$

Minimum Dissipation Conditions of the Mass Transfer and Optimal Separation Sequence
Selection for Multicomponent Mixtures I

117

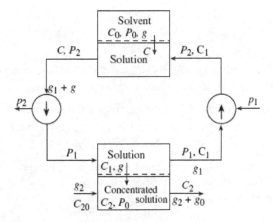

**Figure 14.** Schematic diagram of a diffusion engine with an alternate contact between the working body and sources.

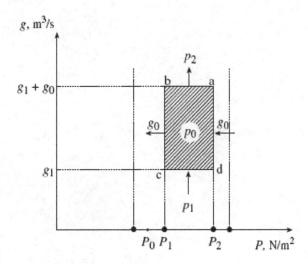

**Figure 15.** Variation cycle for the parameters of the working body in a diffusion engine.

where

The efficiency will be defined as the ratio of power $p$ to the flow rate $g_2$ of the dissolved component:

$$h = \frac{p}{g_2} = \frac{g}{g_2} \Delta P_{21}.$$

The rate of mass transfer is determined by the relations:

$$g = \alpha_1(P_0 + \pi - P_2) = \alpha_1(\pi - \Delta P_{20})$$
$$= \alpha_2[(P_1 + \pi_2) - (P_0 + \pi_1)] = \alpha_2(\Delta \pi_{21} + \Delta P_{10}), \tag{153}$$

where $\Delta P_{20} = P_2 - P_0$, $\Delta \pi_{21} = \pi_2 - \pi_1$, $\Delta P_{10} = P_1 - P_0$. Equation (153) corresponds to the condition that the mass of the working body averaged over the cycle is constant.

Figure 15 demonstrates the cycle of the working body of this diffusion engine. The power $p_1$ is equal to the area of the rectangular $P_2 dc P_1$, and the power $p_2$ to the area of $P_2 ab P_1$. The engine power p is equal to the area of the hatched rectangular $abcd$.

The power of the diffusion engine will be determined when the relationship between the osmotic pressures in the chambers and the flow rates is ignored. To do it, we will solve the problem of constrained optimization:

$$p = (P_2 - P_1)g \to \max_{P_1, P_2}$$

with the constraints:

$$\alpha_1(P_0 + \pi - P_2) = \alpha_2(P_1 - P_0 + \pi_2 - \pi_1) = g. \tag{154}$$

It follows from Eq. (154) that

$$P_1 = \frac{g}{\alpha_2} + P_0 + \pi_1 - \pi_2, \; P_2 = P_0 + \pi - \frac{g}{\alpha_1}.$$

Let us introduce the equivalent permeability:

$$\bar{\alpha} = \frac{\alpha_1 \alpha_2}{\alpha_1 + \alpha_2}$$

and write the equation:

$$P_2 - P_1 = \pi - \pi_1 + \pi_2 - \frac{g}{\alpha}.$$

Then

$$p = g(\pi - \pi_1 + \pi_2 - \frac{g}{\alpha}) = g(\pi + \Delta \pi_{21} - \frac{g}{\alpha}) \to \max_g. \tag{155}$$

The maximum of this expression, which is equal to

$$p^* = \frac{\bar{\alpha}(\pi - \pi_1 + \pi_2)^2}{4} = \frac{\alpha(\pi + \Delta \pi_{21})^2}{4},$$

is reached at

Minimum Dissipation Conditions of the Mass Transfer and Optimal Separation Sequence
Selection for Multicomponent Mixtures I

119

$$g^* = \frac{\bar{\alpha}(\pi - \pi_1 + \pi_2)}{2} = \frac{\alpha(\pi + \Delta\pi_{21})}{2}.$$

Keeping in mind that the osmotic pressures in the chambers are related to the concentrations by Van't Hoff equation (144) and the concentrations are related to the flow rates $g_1$, $g_2$, and $g$, we obtain

$$\pi = CRT = C_1 \frac{g_1 RT}{g_1 + g},$$

$$\Delta\pi_{21} = (C_2 - C_1)RT = \left(\frac{g_2 C_{20} + g C_1}{g_2 + g} - C_1\right)RT.$$

In view of these relations, expression (155) for the engine power takes the form:

$$\begin{aligned}
p &= g\left[RT\left(\frac{C_1 g_1}{g_1 + g} + \frac{g_2 C_{20} + g C_1}{g_2 + g} - C_1\right) - \frac{g}{\alpha}\right] \\
&= g\left[RT\left(\frac{g_2 C_{20} + g C_1}{g_2 + g} - \frac{C_1 g}{g_1 + g}\right) - \frac{g}{\alpha}\right] \to \max_g.
\end{aligned} \tag{156}$$

The expression for the efficiency is written as

$$\eta = \frac{g}{g_2}\left[RT\left(\frac{g_2 C_{20} + g C_1}{g_2 + g} - \frac{C_1 g}{g_1 + g}\right) - \frac{g}{\alpha}\right] \to \max_g. \tag{157}$$

The points of maximum with respect to g for the criteria (156) and (157) coincide. Therefore, we can use either of them in the conditions of optimality to find $g^*$. The stationarity condition of $p$ with respect to $g$ leads to an equation for the optimal flow rate:

$$g = \frac{\bar{\alpha}RT}{2}\left[\left(\frac{g_2^2 C_{20} + 2g g_2 C_1 + g^2 C_1}{(g_2 + g)^2}\right) - C_1 \frac{g(g + 2g_1)}{(g_1 + g)^2}\right]. \tag{158}$$

The solution to Eq. (158) will be $g^*$: it is the optimal value of flow rate g at which the efficiency $\eta$ and power $p$ take their maximal values. The values of flow rate $g^*$ depends on the values of $g_1$, $g_2$, and $C_1$. Its substitution into the equations for $p$ and $\eta$ determines the maximal power $p^*(g_1, g_2, C_1)$ and efficiency $\eta^*(g_1, g_2, C_1)$. The nonnegative nature of $p^*$ and $\eta^*$ imposes constraints on the possible values of $g_1$, $g_2$, and $C_1$. For example, increasing $g_1$ and $g_2$ or decreasing $C_1$ causes an increase in $p^*$.

### 4.4. Conclusion

The estimates obtained in the present paper for the limiting performance of diffusion engines can be used to make their reversible-thermodynamics analysis more accurate and consider the influence of the kinetic factors (mass transfer relations, membrane permeabilities) and production flow rate. These estimates can also be used for the optimization of more complex membrane systems. The capacity of membrane systems increases in proportion to the membrane permeability. In this case, the performance of membranes is decreased by the nonuniformity of concentrations in the solution, polarization phenomena, and the other factors ignored in obtaining the above estimates.

## Author details

A.M. Tsirlin* and I.N. Grigorevsky

*Address all correspondence to: tsirlin@sarc.botik.ru; ivan_ger@mail.ru

The Program Systems Institute of RAS, Pereslavl – Zalessky, Russia

## References

[1] Amelkin, S. A, Burtcler, I. M, Hoffman, K. H, & Tsirlin, A. M. *Estimates of Limiting Possibilities of Separation Process;*. Theor. Found. Chem. Tech. (1983).

[2] Andresen, B. *Finite-Time Thermodynamics*. Copenhagen: Univers, (1983).

[3] Berry, R. S, Kazakov, V, Sieniutycz, S, Szwast, Z, & Tsirlin, A. M. *Thermodynamic Optimization of Finite Time Processes*, Wiley: Chichester, U.K., (1999).

[4] Bosnajakovic, F. *Technical Thermodynamics*; Holt, Rinehart and Winston: New York, (1965). , 2

[5] (Brock, T.D. Membrane Filtration. A User's Guide and Reference Manual, Heidelberg: Seiten-Springer, 1983. Translated under the title Membrannaya fil'tratsiya, Moscow: Mir, 1987).

[6] Curzon, F. L, & Ahlborn, B. *Efficiency of a Carnot Engine at Maximum Power Output*. Am. J. Phys. (1975). , 1975(43), 22-24.

[7]  Dubyaga, V. P, Perepechkin, L. P, & Katalevskii, E. E. *Polymer Membranes*, Moscow: Khimiya, (1981).

[8]  DytnerskiiYu.I., Brykov, V.P., and Kagramanov, G.G., *Membrannoe razdelenie gazov (Membrane Separation of Gases)*, Moscow: Khimiya, (1991).

[9]  Hwang, S, & Kammermeyer, T. K., *Membranes in Separations*, New York: John Wiley & Sons, (1975).

[10] Kesting, R. E. *Synthetic Polymeric Membranes*, New York: Wiley, (1985).

[11] KirshYu.E., *Water Soluble Poly-N-Vinylamides. Synthesis and Physicochemical Properties*, Chichester: Wiley, (1998).

[12] Novikov, I. I. *The Efficiency of Atomic Power Stations*. J. Nucl. Energy II (USSR) (1958). , 1958(7), 125-128.

[13] Orlov, V. N, & Rozonoer, L. I. *Estimates of Efficiency of Controlled Thermodynamic Processes Based on the Balance Equations of Mass, Energy and Entropy*. Abstracts of the 10th All-Union Conference on Control Problems, Moscow, AN SSSR, IPU, (1986). , 75-76.

[14] Orlov, V. N, & Rudenko, A. V. *Optimal Control in Problems of the Limiting Performance of Irreversible Thermodynamic Processes*, Avtom. Telemekh., (1985). (5), 27-41.

[15] Prigogine, I, & Defay, R. *Chemical Thermodynamics*, London: Longmans Green, (1954).

[16] Rozonoer, L. I, & Tsirlin, A. M. *Optimal Control of Thermodynamic Processes*. Autom. Remote Control (1983).

[17] Rozonoer, L. I. *Optimal Thermodynamic Processes with Chemical Reaction and Diffusion, in Thermodynamics and Regulation of Biological processe*; Moscow: Nauka, (1984).

[18] Salamon, P, & Nitzan, A. *Finite Time Optimizations of a Newton's Law Carnot Cycle*, J. Chem. Phys., (1981). , 20(1), 51.

[19] Shambadal, P. *Development and Application of the Concept of Entropy*; Nauka: Moscow, (1967).

[20] Timashev, S. F. *Physical Chemistry of Membrane Processes*, Chichester: Ellis Horwood, (1991).

[21] Tsirlin, A. M. *Irreversible-Thermodynamics Estimates of the Limiting Performance of Thermodynamic and Microeconomic System*; Moscow: Nauka, (2003).

[22] Tsirlin, A. M. *Methods of Averaged Optimization and Their Applications*; Moscow: Fizmatlit, (1997).

[23] Tsirlin, A. M. *Optimal Processes in Thermodynamics and Microeconomics*; Nauka: Moscow, (2003).

[24]  Tsirlin, A. M, Mironova, V. A, Amelkin, S. A, & Kazakov, V. A. *Finite-Time Thermody-namics: Conditions of Minimal Dissipation for Thermodynamic Process with Given Rate.* Phys. ReV. E (1998).

[25]  Tsirlin, A. M. *Separation of Ideal Mixtures in Multistage Systems: An Algorithm for Select-ing a Separation Sequence;* Theoretical Foundations of Chemical Engineering, (2012). , 46(2), 128-134.

[26]  Tsirlin, A. M, & Grigorevsky, I. N. *Thermodynamical estimation of the limit potentialiies of irreversible binary distillation.* J. Non-Equilib. Thermodyn. 35 (2010). , 213-233.

# Chapter 5

# Cross-Diffusion, Viscous Dissipation and Radiation Effects on an Exponentially Stretching Surface in Porous Media

Ahmed A. Khidir and Precious Sibanda

Additional information is available at the end of the chapter

## 1. Introduction

In the last few decades, fluid flow with heat and mass transfer on a continuously stretching surface has attracted considerable attention because of its many applications in industrial and manufacturing processes. Examples of these applications include the drawing of plastic films, glass-fibre and paper production, hot rolling and continuous casting of metals and spinning of fibers. The kinematics of stretching and the simultaneous heating or cooling during such processes play an important role on the structure and quality of the final product.

Sakiadis [30, 31] was the first to study the boundary layer flow due to a continuous moving solid surface. Subsequently, a huge number of studies dealing with different types of fluids, different forms of stretching velocity and temperature distributions have appeared in the literature. Ali [2] investigated similarity solutions of laminar boundary-layer equations in a quiescent fluid driven by a stretched sheet subject to fluid suction or injection. Elbashbeshy [13] extended this problem to a three dimensional exponentially continuous stretching surface. The problem of an exponentially stretching surface with an exponential temperature distribution has been discussed by Magyari and Keller [19]. The problem of mixed convection from an exponentially stretching surface was studied by Partha et al. [24]. They considered the effect of buoyancy and viscous dissipation in the porous medium. They observed that these had a significant effect on the skin friction and the rate of heat transfer. This problem has been extended by Sajid and Hayat [28] who investigated heat transfer over an exponentially stretching sheet in the presence of heat radiation. The same problem was solved numerically by Bidin and Nazar [6] using the Keller-box method. Flow and heat transfer along an exponentially stretching continuous surface with an exponential temperature distribution and an applied magnetic field has been investigated numerically by

Al-Odat et al. [1] while Khan [17] and Sanjayanand and Khan [29] investigated heat transfer due to an exponentially stretching sheet in a viscous-elastic fluid.

Thermal-diffusion and diffusion-thermo effects in boundary layer flow due to a vertical stretching surface have been studied by, *inter alia*, Dursunkaya and Worek [10] while MHD effects, injection/suction, heat radiation, Soret and Dufour effects on the heat and mass transfer on a continuously stretching permeable surface was investigated by El-Aziz [12]. He showed that the Soret and Dufour numbers have a significant influence on the velocity, temperature and concentration distributions.

Srinivasacharya and RamReddy [33] analyzed the problem of mixed convection in a viscous fluid over an exponentially stretching vertical surface subject to Soret and Dufour effects. Ishak [15] investigated the effect of radiation on magnetohydrodynamic boundary layer flow of a viscous fluid over an exponentially stretching sheet. Pal [9] analyzed the effects of magnetic field, viscous dissipation and internal heat generation/absorption on mixed convection heat transfer in the boundary layers on an exponentially stretching continuous surface with an exponential temperature distribution. Loganathan et al. [18] investigated the effect of a chemical reaction on unsteady free convection flow past a semi-infinite vertical plate with variable viscosity and thermal conductivity. They assumed that the viscosity of the fluid was an exponential function and that the thermal conductivity was a linear function of the temperature. They noted that in the case of variable fluid properties, the results obtained differed significantly from those of constant fluid properties. Javed et al. [16] investigated the non-similar boundary layer flow over an exponentially stretching continuous in rotating flow. They observed a reduction in the boundary layer thickness and an enhanced drag force at the surface with increasing fluid rotation.

The aim of the present study is to investigate the effects of cross-diffusion, chemical reaction, heat radiation and viscous dissipation on an exponentially stretching surface subject to an external magnetic field. The wall temperature, solute concentration and stretching velocity are assumed to be exponentially increasing functions. The successive linearisation method (SLM) which has been used in a limited number of studies (see [3, 5, 20–22, 32]) is used to solve the governing coupled non-linear system of equations. Recent studies such as [4, 22, 23] have suggested that the successive linearisation method is accurate and converges rapidly to the numerical results when compared to other semi-analytical methods such as the Adomian decomposition method, the variational iteration method and the homotopy perturbation method. The SLM method can be used in place of traditional numerical methods such as finite differences, Runge-Kutta shooting methods, finite elements in solving non-linear boundary value problems. We compared the results with the Matlab bvp4c numerical routine.

## 2. Governing equations

Consider a quiescent incompressible conducting fluid of constant ambient temperature $T_\infty$ and concentration $C_\infty$ in a porous medium through which an impermeable vertical sheet is stretched with velocity $u_w(x) = u_0 e^{x/\ell}$, temperature distribution $T_w(x) = T_\infty + T_0 e^{2x/\ell}$ and concentration distribution $C_w(x) = C_\infty + C_0 e^{2x/\ell}$ where $C_0$, $T_0$, $u_0$ and $\ell$ are positive constants. The $x$-axis is directed along the continuous stretching surface and the $y$-axis is normal to the surface. A variable magnetic field $B(x)$ is applied in the $y$-direction. In

addition, heat radiation and cross-diffusion effects are considered to be significant. The
governing boundary-layer equations subject to the Boussinesq approximations are

$$\frac{\partial u}{\partial x} + \frac{\partial v}{\partial y} = 0, \tag{1}$$

$$u\frac{\partial u}{\partial x} + v\frac{\partial u}{\partial y} = v\frac{\partial^2 u}{\partial y^2} + g\beta_T(T - T_\infty) + g\beta_C(C - C_\infty) - \left(\frac{v}{K} + \frac{\sigma B^2}{\rho}\right)u, \tag{2}$$

$$u\frac{\partial T}{\partial x} + v\frac{\partial T}{\partial y} = \frac{k}{\rho c_p}\frac{\partial^2 T}{\partial y^2} + \frac{v}{c_p}\left(\frac{\partial u}{\partial y}\right)^2 + \frac{D_m K_T}{c_s c_p}\frac{\partial^2 C}{\partial y^2} - \frac{1}{\rho c_p}\frac{\partial q_r}{\partial y}, \tag{3}$$

$$u\frac{\partial C}{\partial x} + v\frac{\partial C}{\partial y} = D_m\frac{\partial^2 C}{\partial y^2} + \frac{D_m K_T}{T_m}\frac{\partial^2 T}{\partial y^2} - \gamma(C - C_\infty), \tag{4}$$

The boundary conditions are given by

$$\left.\begin{array}{l} u = u_w(x), v = 0, T = T_w(x), C = C_w(x) \quad \text{at} \quad y = 0, \\ u \to 0, T \to T_\infty, C \to C_\infty \quad \text{as} \quad y \to \infty. \end{array}\right\} \tag{5}$$

where $u$ and $v$ are the velocity components along the $x$ and $y$ axis, respectively, $T$ and $C$
denote the temperature and concentration, respectively, $K$ is the permeability of the porous
medium, $v$ is the kinematic viscosity, $g$ is the acceleration due to gravity, $\beta_T$ is the coefficient
of thermal expansion, $\beta_C$ is the coefficient of concentration expansion, $B$ is the uniform
magnetic field, $\rho$ is the liquid density, $\sigma$ is the electrical conductivity, $D_m$ is the mass
diffusivity, $c_s$ is the concentration susceptibility, $c_p$ is the specific heat capacity, $T_m$ is the
mean fluid temperature, $K_T$ is the thermal diffusion ratio and $\gamma$ is the rate of chemical
reaction.

The radiative heat flux term $q_r$ is given by the Rosseland approximation (see Raptis [26] and
Sparrow [27]);

$$q_r = -\frac{4\sigma^*}{3k^*}\frac{\partial T^4}{\partial y}, \tag{6}$$

where $\sigma^*$ and $k^*$ are the Stefan-Boltzman constant and the mean absorption coefficient,
respectively. We assume that the term $T^4$ may be expanded in a Taylor series about $T_\infty$
and neglecting higher-order terms to get

$$T^4 \cong 4T_\infty^3 T - 3T_\infty^4, \tag{7}$$

Substituting equations (6) and (7) in equation (3) gives

$$u\frac{\partial T}{\partial x} + v\frac{\partial T}{\partial y} = \left(\frac{k}{\rho c_p} + \frac{16\sigma^* T_\infty^3}{3\rho c_p k^*}\right)\frac{\partial^2 T}{\partial y^2} + \frac{v}{c_p}\left(\frac{\partial u}{\partial y}\right)^2 + \frac{D_m K_T}{c_s c_p}\frac{\partial^2 C}{\partial y^2}, \tag{8}$$

A similarity solutions may be obtained by assuming that the magnetic field term $B(x)$ has the form

$$B(x) = B_0 e^{x/2\ell} \tag{9}$$

where $B_0$ is the constant magnetic field. The system of partial differential equations (1) - (4) and (8) can be simplified further by introducing the stream function $\psi$ where

$$u = \frac{\partial \psi}{\partial y} \quad \text{and} \quad v = -\frac{\partial \psi}{\partial x}, \tag{10}$$

together with transformations

$$\left. \begin{array}{l} \eta = \frac{y}{L}\sqrt{\frac{Re}{2}}\, e^{x/2\ell}, \psi = \sqrt{2Re}v\, e^{x/2\ell} f(\eta), \\ T - T_\infty + T_0 e^{2x/\ell}\theta(\eta), C = C_\infty + C_0 e^{2x/\ell}\phi(\eta) \end{array} \right\}. \tag{11}$$

Substituting (11) into the governing partial differential equations gives

$$f''' + ff'' - 2f'^2 - \left(M + \frac{1}{Re_D}\right)f' + 2\frac{Gr_x}{Re^2}(\theta + N_1\phi) = 0, \tag{12}$$

$$\frac{1}{Pr}\left(1 + \frac{4}{3}R_d\right)\theta'' + f\theta' - 4f'\theta + Gb(f'')^2 + D_f\phi'' = 0, \tag{13}$$

$$\frac{1}{Sc}\phi'' + f\phi' - 4f'\phi + Sr\theta'' - 2R\phi = 0. \tag{14}$$

The corresponding dimensionless boundary conditions take the form

$$\left. \begin{array}{l} f(\eta) = 0,\ f'(\eta) = 1,\ \theta(\eta) = 1,\ \phi(\eta) = 1 \quad \text{at} \quad \eta = 0 \\ f'(\eta) \to 0,\ \theta(\eta) \to 0,\ \phi(\eta) \to 0 \quad \text{as} \quad \eta \to \infty \end{array} \right\} \tag{15}$$

where $M$ is the magnetic parameter, $Gr_x$ is the Grashof number $Re$ is the Reynolds number, $N_1$ is the buoyancy ratio, $Re_D$ is the Darcy-Reynolds number, $Da$ is the Darcy number, $Pr$ is the Prandtl number, $R_d$ is the thermal radiation parameter, $Gb$ is the viscous dissipation parameter or Gebhart number, $D_f$ is the Dufour number, $Sc$ is the Schmidt number, $S_r$ is the Soret number and $R$ is the chemical reaction rate parameter. These parameters are defined as

$$M = \frac{2\sigma B_0^2 \ell}{\rho u_0}, \quad Gr_x = \frac{g\beta_T T_0 \ell^3 e^{2x/\ell}}{v^2}, \quad Re = \frac{u_w \ell}{v}, \quad N_1 = \frac{\beta_c C_0}{\beta_T T_0}, \tag{16}$$

$$Re_D = \frac{2}{ReDa}, \quad Da = \frac{K}{\ell^2}, \quad Pr = \frac{v}{\alpha}, \quad R_d = \frac{4\sigma^* T_\infty^3}{kk^*}, \quad Gb = \frac{u_0^2}{c_p T_0}, \tag{17}$$

$$D_f = \frac{D_m K_T C_0}{c_s c_p v T_0}, \quad Sc = \frac{v}{D_m}, \quad S_r = \frac{D_m K_T T_0}{T_m v C_0}, \quad R = \frac{\alpha \ell}{u_0}. \tag{18}$$

The ratio $Gr_x/Re^2$ in equation (12) is the mixed convection parameter which represents aiding buoyancy if $Gr_x/Re^2 > 0$ and opposing buoyancy if $Gr_x/Re^2 < 0$. The skin friction coefficient $C_{fx}$, the Nusselt number $Nu_x$ and the Sherwood $Sh_x$ number are given by

$$C_{fx} = \frac{2\mu}{\rho u_w^2} \frac{\partial u}{\partial y}\bigg|_{y=0} = \sqrt{\frac{2x}{\ell Re_x}}\, f''(0), \tag{19}$$

$$Nu_x = -\frac{x}{T_w - T_\infty} \frac{\partial T}{\partial y}\bigg|_{y=0} = -\sqrt{\frac{x Re_x}{2\ell}}\, \theta'(0) \tag{20}$$

$$Sh_x = -\frac{x}{C_w - C_\infty} \frac{\partial C}{\partial y}\bigg|_{y=0} = -\sqrt{\frac{x Re_x}{2\ell}}\, \phi'(0) \tag{21}$$

where $Re_x = x u_w(x)/v$ is the local Reynolds number.

## 3. Method of solution

The system of equations (12)-(14) together with the boundary conditions (15) were solved using a successive linearisation method (SLM) (see [22, 32]). The SLM is based on the assumption that the unknown functions $f(\eta)$, $\theta(\eta)$ and $\phi(\eta)$ can be expanded as

$$f(\eta) = f_i(\eta) + \sum_{m=0}^{i-1} F_m(\eta), \theta(\eta) = \theta_i(\eta) + \sum_{m=0}^{i-1} \Theta_m(\eta), \phi(\eta) = \phi_i(\eta) + \sum_{m=0}^{i-1} \Phi_m(\eta), \tag{22}$$

where $f_i$, $\theta_i$ and $\phi_i$ are unknown functions and $F_m$, $\Theta_m$ and $\Phi_m$ ($m \geq 1$) are successive approximations which are obtained by recursively solving the linear part of the equation system that results from substituting firstly expansions in the governing equations. The initial guesses $F_0(\eta)$, $\Theta_0(\eta)$ and $\Phi_0(\eta)$ are chosen to satisfy the boundary condition

$$\left.\begin{array}{l} F_0(\eta) = 0,\ F_0'(\eta) = 1,\ \Theta_0(\eta) = 1,\ \Phi_0(\eta) = 1 \quad \text{at} \quad \eta = 0 \\ F_0'(\eta) \to 0,\ \Theta_0(\eta) \to 0,\ \Phi_0(\eta) \to 0 \quad \text{as} \quad \eta \to \infty \end{array}\right\}. \tag{23}$$

Suitable choices in this problem are

$$F_0(\eta) = 1 - e^{-\eta},\ \Theta_0(\eta) = e^{-\eta} \quad \text{and} \quad \Phi_0(\eta) = e^{-\eta}. \tag{24}$$

Starting from the initial guesses, the subsequent solutions $F_i$, $\Theta_i$ and $\Phi_i$ ($i \geq 1$) are obtained by successively solving the linearised form of the equations which are obtained by substituting equation (22) in the governing equations. The linearised equations to be solved are

$$a_{1,i-1}F_i''' + a_{2,i-1}F_i'' + a_{3,i-1}F_i' + a_{4,i-1}F_i + a_{5,i-1}\Theta_i + a_{6,i-1}\Phi_i = r_{1,i-1}, \tag{25}$$

$$b_{1,i-1}\Theta_i'' + b_{2,i-1}\Theta_i' + b_{3,i-1}\Theta_i + b_{4,i-1}F_i'' + b_{5,i-1}F_i' + b_{6,i-1}F_i + b_{7,i-1}\Phi_i = r_{2,i-1}, \tag{26}$$

$$c_{1,i-1}\Phi_i'' + c_{2,i-1}\Phi_i' + c_{3,i-1}\Phi_i + c_{4,i-1}F_i' + c_{5,i-1}F_i + c_{6,i-1}\Theta_i'' = r_{3,i-1}. \tag{27}$$

subject to the boundary conditions

$$F_i(0) = F_i'(0) = F_i'(\infty) = \Theta_i(0) = \Theta_i(\infty) = \Phi_i(0) = \Phi_i(\infty) = 0, \tag{28}$$

where the coefficient parameters are

$$a_{1,i-1} = -1, \quad a_{2,i-1} = \sum_{m=0}^{i-1} f_m', \quad a_{3,i-1} = -4 \sum_{m=0}^{i-1} f_m' - M - \frac{1}{Re_D}, a_{4,i-1} = \sum_{m=0}^{i-1} f_m''$$

$$a_{5,i-1} = 2\frac{Gr_x}{Re^2}, \, a_{6,i-1} = 2N_1 \frac{Gr_x}{Re^2}, \, b_{1,i-1} = \sum_{m=0}^{i-1} f_m, \, b_{2,i-1} = -4 \sum_{m=0}^{i-1} \theta_m, b_{3,i-1} = \sum_{m=0}^{i-1} \theta_m',$$

$$b_{4,i-1} = \frac{3+4R_d}{3Pr}, \, b_{5,i-1} = \sum_{m=0}^{i-1} f_m, \, b_{6,i-1} = -4 \sum_{m=0}^{i-1} f_m', \, b_{7,i-1} = D_f, \, c_{1,i-1} = -4 \sum_{m=0}^{i-1} \phi_m$$

$$c_{2,i-1} = \sum_{m=0}^{i-1} \phi_m', \, c_{3,i-1} = Sr, \, c_{4,i-1} = \frac{1}{Sc}, \, c_{5,i-1} = \sum_{m=0}^{i-1} f_m, \, c_{6,i-1} = \quad 2R - 4 \sum_{m=0}^{i-1} f_m',$$

$$r_{1,i-1} = -\sum_{m=0}^{i-1} f_m''' - \sum_{m=0}^{i-1} f_m \sum_{m=0}^{i-1} f_m'' + 2 \sum_{m=0}^{i-1} f_m'^2 + \left( M + \frac{1}{Re_D} \right) \sum_{m=0}^{i-1} f_m'$$

$$-\frac{2Gr_x}{Re^2} \sum_{m=0}^{i-1} (\theta_m + N_1 \phi_m)$$

$$r_{2,i-1} = -\sum_{m=0}^{i-1} \frac{1}{Pr}(\phi_m'' + \frac{4R_d}{3Pr}\theta_m'') - \sum_{m=0}^{i-1} f_m \sum_{m=0}^{i-1} \theta_m' + 4 \sum_{m=0}^{i-1} f_m' \sum_{m=0}^{i-1} \theta_m - Gb \sum_{m=0}^{i-1} f_m''^2 -$$

$$D_f \sum_{m=0}^{i-1} \phi_m''$$

$$r_{3,i-1} = -\frac{1}{Sc} \sum_{m=0}^{i-1} \phi_m'' \sum_{m=0}^{i-1} f_m \sum_{m=0}^{i-1} \phi_m' + 4 \sum_{m=0}^{i-1} f_m \sum_{m=0}^{i-1} \theta_m - Sr \sum_{m=0}^{i-1} \phi_m'' + 2R \sum_{m=0}^{i-1} \phi_m$$

The solutions $F_i$, $\Theta_i$ and $\Phi_i$ for $i \geq 1$ are found by iteratively solving equations (25)-(27). Finally, after $M$ iterations, the solutions $f(\eta)$, $\theta(\eta)$ and $\phi(\eta)$ may be written as

$$f(\eta) \approx \sum_{m=0}^{M} F_m(\eta), \theta(\eta) \approx \sum_{m=0}^{M} \Theta_m(\eta), \Phi(\eta) \approx \sum_{m=0}^{M} \Phi_m(\eta). \tag{29}$$

where $M$ is termed the order of SLM approximation. Equations (25)-(27) are solved using the Chebyshev spectral collocation method. We first transform the domain of solution $[0, \infty)$ into the domain $[-1, 1]$ using the domain truncation technique where the problem is solved in the interval $[0, L]$ where $L$ is a scaling parameter used to invoke the boundary condition at infinity. This is achieved by using the mapping

$$\frac{\eta}{L} = \frac{\xi + 1}{2}, -1 \le \xi \le 1, \tag{30}$$

We discretize the domain $[-1, 1]$ using the Gauss-Lobatto collocation points given by

$$\xi = \cos \frac{\pi j}{N}, j = 0, 1, 2, \dots, N, \tag{31}$$

where $N$ is the number of collocation points used. The functions $F_i$, $\Theta_i$ and $\Phi_i$ for $i \ge 1$ are approximated at the collocation points as follows

$$F_i(\xi) \approx \sum_{k=0}^{N} F_i(\xi_k) T_k(\xi_j), \Theta_i(\xi) \approx \sum_{k=0}^{N} \Theta_i(\xi_k) T_k(\xi_j), \Phi_i(\xi) \approx \sum_{k=0}^{N} \Phi_i(\xi_k) T_k(\xi_j) j = 0, 1, \dots, N,$$
$$\tag{32}$$

where $T_k$ is the $k^{\text{th}}$ Chebyshev polynomial given by

$$T_k(\xi) = \cos \left[ k \cos^{-1}(\xi) \right]. \tag{33}$$

The derivatives of the variables at the collocation points are represented as

$$\frac{d^r F_i}{d\eta^r} = \sum_{k=0}^{N} \mathbf{D}_{kj}^r F_i(\xi_k), \frac{d^r \Theta_i}{d\eta^r} = \sum_{k=0}^{N} \mathbf{D}_{kj}^r \Theta_i(\xi_k), \frac{d^r \Phi_i}{d\eta^r} = \sum_{k=0}^{N} \mathbf{D}_{kj}^r \Phi_i(\xi_k) j = 0, 1, \dots, N, \tag{34}$$

where $r$ is the order of differentiation and $\mathbf{D} = \frac{2}{L}\mathcal{D}$ with $\mathcal{D}$ being the Chebyshev spectral differentiation matrix (see, for example [7, 8]), whose entries are defined as

$$\left. \begin{aligned} \mathcal{D}_{00} &= \frac{2N^2 + 1}{6}, \\ \mathcal{D}_{jk} &= \frac{c_j}{c_k} \frac{(-1)^{j+k}}{\xi_j - \xi_k}, j \ne k; j, k = 0, 1, \dots, N, \\ \mathcal{D}_{kk} &= -\frac{\xi_k}{2(1 - \xi_k^2)}, k = 1, 2, \dots, N - 1, \\ \mathcal{D}_{NN} &= -\frac{2N^2 + 1}{6}. \end{aligned} \right\} \tag{35}$$

Substituting equations (30)-(34) into equations (25)-(27) leads to the matrix equation

$$\mathbf{A}_{i-1} \mathbf{X}_i = \mathbf{R}_{i-1}, \tag{36}$$

In equation (36), $\mathbf{A}_{i-1}$ is a $(3N+3) \times (3N+3)$ square matrix and $\mathbf{X}_i$ and $\mathbf{R}_{i-1}$ are $(3N+3) \times 1$ column vectors defined by

$$\mathbf{A}_{i-1} = \begin{bmatrix} A_{11} & A_{12} & A_{13} \\ A_{21} & A_{22} & A_{23} \\ A_{31} & A_{32} & A_{33} \end{bmatrix}, \mathbf{X}_i = \begin{bmatrix} F_i \\ \Theta_i \\ \Phi_i \end{bmatrix}, \mathbf{R}_{i-1} = \begin{bmatrix} \mathbf{r}_{1,i-1} \\ \mathbf{r}_{2,i-1} \\ \mathbf{r}_{3,i-1} \end{bmatrix}, \tag{37}$$

where

$$F_i = [f_i(\xi_0), f_i(\xi_1), ..., f_i(\xi_{N-1}), f_i(\xi_N)]^T,$$

$$\Theta_i = [\theta_i(\xi_0), \theta_i(\xi_1), ..., \theta_i(\xi_{N-1}), \theta_i(\xi_N)]^T,$$

$$\Phi_i = [\phi_i(\xi_0), \phi_i(\xi_1), ..., \phi_i(\xi_{N-1}), \phi_i(\xi_N)]^T,$$

$$\mathbf{r}_{1,i-1} = [r_{1,i-1}(\xi_0), r_{1,i-1}(\xi_1), ..., r_{1,i-1}(\xi_{N-1}), r_{1,i-1}(\xi_N)]^T,$$

$$\mathbf{r}_{2,i-1} = [r_{2,i-1}(\xi_0), r_{2,i-1}(\xi_1), ..., r_{2,i-1}(\xi_{N-1}), r_{2,i-1}(\xi_N)]^T,$$

$$\mathbf{r}_{3,i-1} = [r_{3,i-1}(\xi_0), r_{3,i-1}(\xi_1), ..., r_{3,i-1}(\xi_{N-1}), r_{3,i-1}(\xi_N)]^T,$$

$$A_{11} = \mathbf{a}_{1,i-1}\mathbf{D}^3 + \mathbf{a}_{2,i-1}\mathbf{D}^2 + \mathbf{a}_{3,i-1}\mathbf{D} + \mathbf{a}_{4,i-1}\mathbf{I}, A_{12} = \mathbf{a}_{5,i-1}\mathbf{I} + \mathbf{a}_{6,i-1}\mathbf{I}, A_{13} = \mathbf{I},$$

$$A_{21} = \mathbf{b}_{1,i-1}\mathbf{D}^2 + \mathbf{b}_{2,i-1}\mathbf{D} + \mathbf{b}_{3,i-1}\mathbf{I}, A_{22} = \mathbf{b}_{4,i-1}\mathbf{D}^2 + \mathbf{b}_{5,i-1}\mathbf{D} + \mathbf{b}_{6,i-1}\mathbf{I}, A_{23} = \mathbf{b}_{7,i-1}\mathbf{D}^2,$$

$$A_{31} = \mathbf{c}_{1,i-1}\mathbf{D}^2 + \mathbf{c}_{2,i-1}\mathbf{D} + \mathbf{c}_{3,i-1}\mathbf{I}, A_{32} = \mathbf{c}_{4,i-1}\mathbf{D} + \mathbf{c}_{5,i-1}\mathbf{I}, A_{33} = \mathbf{c}_{6,i-1}\mathbf{D}^2.$$

In the above definitions $T$ stands for transpose, $\mathbf{a}_{k,i-1}$ $(k = 1,...,6)$, $\mathbf{b}_{k,i-1}$ $(k = 1,...,7)$, $\mathbf{c}_{k,i-1}$ $(k = 1,...,6)$, and $\mathbf{r}_{k,i-1}$ $(k = 1,2,3)$ are diagonal matrices of order $(N+1) \times (N+1)$, $\mathbf{I}$ is an identity matrix of order $(N+1) \times (N+1)$. Finally the solution is obtained as

$$\mathbf{X}_i = \mathbf{A}_{i-1}^{-1}\mathbf{R}_{i-1}. \tag{38}$$

## 4. Results and discussion

In generating the results presented here it was determined through numerical experimentation that $L = 15$ and $N = 60$ gave sufficient accuracy for the linearisation method. In addition, the results in this work were obtained for $Pr = 0.71$ which physically corresponds to air and the Schmidt number $Sc = 0.22$ for hydrogen at approximately $25°$ and one atmospheric pressure. The Darcy-Reynolds number was fixed at $Re_D = 100$.

Tables 1 - 7 show, firstly the effects of various parameters on the skin-friction, the local heat and the mass transfer coefficients for different physical parameters values. Secondly, to confirm the accuracy of the linearisation method, these results are compared to those obtained using the Matlab bvp4c solver. The results from the two methods are in excellent agreement with the linearisation method converging at the four order with accuracy of up to six decimal places.

The effect of increasing the magnetic filed parameter $M$ on the skin-friction coefficient $f''(0)$, the Nusselt number $-\theta'(0)$ and the Sherwood number $-\phi'(0)$ are given in Table 1. Here we find that increasing the magnetic filed parameter leads to reduces Nusselt number and Sherwood number as well as skin friction coefficient in case of aiding buoyancy. These results are to be expected, and are, in fact, similar to those obtained previously by, among others (Ishak [15] and Ibrahim and Makinde [14]).

| | $M$ | SLM results | | | | bvp4c |
|---|---|---|---|---|---|---|
| | | 1st order | 2nd order | 3rd order | 4th order | |
| $f''(0)$ | 0.0 | -0.130330 | -0.137803 | -0.138236 | -0.138242 | -0.138242 |
| | 0.1 | -0.172947 | -0.179731 | -0.179950 | -0.179952 | -0.179952 |
| | 0.5 | -0.334846 | -0.339272 | -0.339242 | -0.339242 | -0.339242 |
| | 1.0 | -0.521003 | -0.523599 | -0.523568 | -0.523568 | -0.523568 |
| $-\theta'(0)$ | 0.0 | 1.422819 | 1.354658 | 1.354263 | 1.354252 | 1.354252 |
| | 0.1 | 1.407222 | 1.345231 | 1.345006 | 1.345004 | 1.345003 |
| | 0.5 | 1.349392 | 1.308530 | 1.308480 | 1.308480 | 1.308480 |
| | 1.0 | 1.286198 | 1.264064 | 1.264037 | 1.264037 | 1.264037 |
| $-\phi'(0)$ | 0.0 | 1.297706 | 1.288178 | 1.288065 | 1.288063 | 1.288063 |
| | 0.1 | 1.294285 | 1.285273 | 1.285212 | 1.285212 | 1.285212 |
| | 0.5 | 1.281653 | 1.274574 | 1.274572 | 1.274572 | 1.274572 |
| | 1.0 | 1.267871 | 1.262761 | 1.262760 | 1.262760 | 1.262760 |

**Table 1.** The effect of various values of $M$ on skin-friction, heat and mass transfer coefficients when $Gr_x/Re^2 = 1.5$, $Gb = 0.5$, $R_d = 0.2$, $D_f = 0.3$, $Sr = 0.2$, $R = 2$ and $N_1 = 0.1$

In Table 2 an increase in the mixed convection parameter $Gr_x/Re^2$ (that is, aiding buoyancy) enhances the skin friction coefficient. This is explained by the fact that an increase in the fluid buoyancy leads to an acceleration of the fluid flow, thus increasing the skin friction coefficient. Similar results were obtained in the past by Srinivasacharya and RamReddy [33] and Partha et al. [24]. Also, the non-dimensional heat and mass transfer coefficients increase when $Gr_x/Re^2$ increases. This is because an increasing in mixed convection parameter, increases the momentum transport in the boundary layer this is leads to carried out more heat and mass species out of the surface, then reducing the thermal and concentration boundary layers thickness and hence increasing the heat and mass transfer rates.

Tables 3 and 4 show the effects of increasing the radiation parameter $R_d$ and the chemical reaction parameter $R$ on the skin-friction, and the heat and mass transfer rates respectively. The skin-friction coefficient is enhanced by the radiation parameter. It is however reduced by the chemical reaction parameter (Loganathan et. al[18]). Increasing the radiation parameter $R_d$ and chemical reaction parameter $R$ have the same effect on heat and mass transfer rates, that is, $-\theta'(0)$ decreases while $-\phi'(0)$ is increases. Large values of $R_d$ and $R$ lead to a decrease in the buoyancy force and, consequently, a decrease in the thicknesses of both the thermal and the momentum boundary layers (see Sajid [28]).

|  | $Gr_x/Re^2$ | SLM results | | | | bvp4c |
|  |  | 1st order | 2nd order | 3rd order | 4th order |  |
|---|---|---|---|---|---|---|
| $f''(0)$ | 0.0 | -1.459148 | -1.469821 | -1.469885 | -1.469885 | -1.469885 |
|  | 0.5 | -1.042194 | -1.044162 | -1.044208 | -1.044208 | -1.044208 |
|  | 1.0 | -0.674522 | -0.678490 | -0.678447 | -0.678447 | -0.678447 |
|  | 1.5 | -0.334846 | -0.339272 | -0.339242 | -0.339242 | -0.339242 |
| $-\theta'(0)$ | 0.0 | 0.973213 | 0.934517 | 0.933372 | 0.933372 | 0.933372 |
|  | 0.5 | 1.135051 | 1.134112 | 1.134083 | 1.134083 | 1.134083 |
|  | 1.0 | 1.253215 | 1.236804 | 1.236738 | 1.236738 | 1.236738 |
|  | 1.5 | 1.349392 | 1.308530 | 1.308480 | 1.308480 | 1.308480 |
| $-\phi'(0)$ | 0.0 | 1.200469 | 1.201715 | 1.201709 | 1.201709 | 1.201709 |
|  | 0.5 | 1.234639 | 1.233314 | 1.233299 | 1.233299 | 1.233299 |
|  | 1.0 | 1.260335 | 1.255659 | 1.255655 | 1.255655 | 1.255655 |
|  | 1.5 | 1.281653 | 1.274574 | 1.274572 | 1.274572 | 1.274572 |

**Table 2.** The effect of various values of $Gr_x/Re^2$ on skin-friction, heat and mass transfer coefficients when $M = 0.5$, $Gb = 0.5$, $R_d = 0.2$, $D_f = 0.3$, $Sr = 0.2$, $R = 2$ and $N_1 = 0.1$

|  | $R_d$ | SLM results | | | | bvp4c |
|  |  | 1st order | 2nd order | 3rd order | 4th order |  |
|---|---|---|---|---|---|---|
| $f''(0)$ | 0.0 | -0.390151 | -0.384799 | -0.385087 | -0.385088 | -0.385088 |
|  | 0.2 | -0.334846 | -0.339272 | -0.339242 | -0.339242 | -0.339242 |
|  | 0.5 | -0.267055 | -0.286928 | -0.286715 | -0.286715 | -0.286715 |
|  | 1.0 | -0.178534 | -0.223773 | -0.224247 | -0.224239 | -0.224239 |
| $-\theta'(0)$ | 0.0 | 1.508929 | 1.466549 | 1.466541 | 1.466540 | 1.466540 |
|  | 0.2 | 1.349392 | 1.308530 | 1.308480 | 1.308480 | 1.308480 |
|  | 0.5 | 1.184709 | 1.146728 | 1.146321 | 1.146321 | 1.146321 |
|  | 1.0 | 1.008715 | 0.976294 | 0.974693 | 0.974674 | 0.974674 |
| $-\phi'(0)$ | 0.0 | 1.267312 | 1.263619 | 1.263567 | 1.263567 | 1.263567 |
|  | 0.2 | 1.281653 | 1.274574 | 1.274572 | 1.274572 | 1.274572 |
|  | 0.5 | 1.298805 | 1.286546 | 1.286483 | 1.286483 | 1.286483 |
|  | 1.0 | 1.320928 | 1.300398 | 1.299833 | 1.299832 | 1.299832 |

**Table 3.** The effect of $R_d$ on skin-friction, heat and mass transfer coefficients when $M = 0.5$, $Gr_x/Re^2 = 1.5$, $Gb = 0.5$, $D_f = 0.3$, $Sr = 0.2$, $R = 2$ and $N_1 = 0.1$

| | $R$ | SLM results | | | | bvp4c |
|---|---|---|---|---|---|---|
| | | 1st order | 2nd order | 3rd order | 4th order | |
| $f''(0)$ | 0.0 | -0.332793 | -0.334754 | -0.334629 | -0.334629 | -0.334629 |
| | 0.5 | -0.333949 | -0.336858 | -0.336797 | -0.336797 | -0.336797 |
| | 1.0 | -0.334431 | -0.337995 | -0.337952 | -0.337952 | -0.337952 |
| | 3.0 | -0.334965 | -0.339959 | -0.339936 | -0.339936 | -0.339936 |
| $-\theta'(0)$ | 0.0 | 1.413844 | 1.364409 | 1.364235 | 1.364232 | 1.364232 |
| | 0.5 | 1.392460 | 1.347513 | 1.347472 | 1.347472 | 1.347472 |
| | 1.0 | 1.376336 | 1.333290 | 1.333247 | 1.333247 | 1.333247 |
| | 3.0 | 1.326247 | 1.286737 | 1.286682 | 1.286682 | 1.286682 |
| $-\phi'(0)$ | 0.0 | 0.773109 | 0.807913 | 0.809416 | 0.809448 | 0.809448 |
| | 0.5 | 0.950211 | 0.956980 | 0.957025 | 0.957025 | 0.957025 |
| | 1.0 | 1.076976 | 1.076044 | 1.076055 | 1.076055 | 1.076055 |
| | 3.0 | 1.452263 | 1.442740 | 1.442734 | 1.442734 | 1.442734 |

**Table 4.** The effect of $R$ on skin-friction, heat and mass transfer coefficients when
$M = 0.5$, $Gr_x/Re^2 = 1.5$, $Gb = 0.5$, $R_d = 0.2$, $D_f = 0.3$, $Sr = 0.2$ and $N_1 = 0.1$

Table 5 shows the influence of the viscous dissipation parameter $Gb$. The skin-friction coefficient and the Sherwood number increase as $Gb$ increases. However, the heat transfer rate is reduced when $Gb$ is increased.

The effect of the Soret parameter on the skin-friction, the heat and the mass transfer coefficients is presented in Table 6. Clearly, increasing this parameter leads to increase in the heat transfer rate and a decrease in both the skin friction coefficient and the mass transfer rate. Similar findings were reported by Partha et al. [25].

Table 7 shows the effect of the Dufour number on the skin-friction, the heat and the mass transfer coefficients. It seen that as the Dufour parameter increases, the skin-friction coefficient and mass transfer rate are enhanced while the mass transfer rate is reduced. The Soret and Dufour numbers have opposite effects on Nusselt and Sherwood numbers.

|         |       | SLM results |           |           |           |           |
|---------|-------|-------------|-----------|-----------|-----------|-----------|
|         | $Gb$  | 1st order   | 2nd order | 3rd order | 4th order | bvp4c     |
| $f''(0)$ | 0.0  | -0.351301   | -0.354490 | -0.354558 | -0.354558 | -0.354558 |
|         | 0.5   | -0.334846   | -0.339272 | -0.339242 | -0.339242 | -0.339242 |
|         | 1.0   | -0.320079   | -0.325029 | -0.324930 | -0.324930 | -0.324930 |
|         | 2.0   | -0.295309   | -0.299203 | -0.298988 | -0.298988 | -0.298988 |
| $-\theta'(0)$ | 0.0 | 1.364554 | 1.341183  | 1.341179  | 1.341179  | 1.341179  |
|         | 0.5   | 1.349392    | 1.308530  | 1.308480  | 1.308480  | 1.308480  |
|         | 1.0   | 1.340317    | 1.278965  | 1.278884  | 1.278884  | 1.278883  |
|         | 2.0   | 1.337606    | 1.227792  | 1.227640  | 1.227640  | 1.227640  |
| $-\phi'(0)$ | 0.0 | 1.278392  | 1.271621  | 1.271607  | 1.271607  | 1.271607  |
|         | 0.5   | 1.281653    | 1.274574  | 1.274572  | 1.274572  | 1.274572  |
|         | 1.0   | 1.284491    | 1.277309  | 1.277310  | 1.277310  | 1.277310  |
|         | 2.0   | 1.289013    | 1.282196  | 1.282194  | 1.282194  | 1.282194  |

**Table 5.** The effect of $Gb$ on skin-friction, heat and mass transfer coefficients when $M = 0.5$, $Gr_x/Re^2 = 1.5$, $R_d = 0.2$, $D_f = 0.3$, $Sr = 0.2$, $R = 2$ and $N_1 = 0.1$

|         |       | SLM results |           |           |           |           |
|---------|-------|-------------|-----------|-----------|-----------|-----------|
|         | $Sr$  | 1st order   | 2nd order | 3rd order | 4th order | bvp4c     |
| $f''(0)$ | 0.1  | -0.271640   | -0.289773 | -0.289559 | -0.289559 | -0.289559 |
|         | 0.5   | -0.332574   | -0.337635 | -0.337603 | -0.337603 | -0.337603 |
|         | 1.0   | -0.338023   | -0.341848 | -0.341863 | -0.341863 | -0.341863 |
|         | 1.5   | -0.338296   | -0.341846 | -0.341878 | -0.341878 | -0.341878 |
| $-\theta'(0)$ | 0.1 | 1.192649 | 1.151536  | 1.151190  | 1.151190  | 1.151190  |
|         | 0.5   | 1.343705    | 1.304565  | 1.304512  | 1.304512  | 1.304512  |
|         | 1.0   | 1.363632    | 1.324409  | 1.324369  | 1.324369  | 1.324369  |
|         | 1.5   | 1.371163    | 1.331565  | 1.331527  | 1.331527  | 1.331527  |
| $-\phi'(0)$ | 0.1 | 1.311432  | 1.298783  | 1.298733  | 1.298733  | 1.298733  |
|         | 0.5   | 1.233004    | 1.228589  | 1.228590  | 1.228590  | 1.228590  |
|         | 1.0   | 1.146766    | 1.147516  | 1.147516  | 1.147516  | 1.147516  |
|         | 1.5   | 1.061229    | 1.066875  | 1.066878  | 1.066878  | 1.066878  |

**Table 6.** The effect of $Sr$ on skin-friction, heat and mass transfer coefficients when $M = 0.5$, $Gr_x/Re^2 = 1.5$, $Gb = 0.5$, $R_d = 0.3$, $R = 2$ and $N_1 = 0.1$

| | $D_f$ | SLM results | | | | bvp4c |
| | | 1st order | 2nd order | 3rd order | 4th order | |
|---|---|---|---|---|---|---|
| | 0.1 | -0.2897226 | -0.3041903 | -0.3039915 | -0.3039916 | -0.3039916 |
| $f''(0)$ | 0.3 | -0.2670553 | -0.2869281 | -0.2867154 | -0.2867154 | -0.2867154 |
| | 0.7 | -0.2181199 | -0.2504226 | -0.2503761 | -0.2503747 | -0.2503747 |
| | 1.5 | -0.1168316 | -0.1785689 | -0.1797718 | -0.1797642 | -0.1797642 |
| | 0.1 | 1.2407545 | 1.2030266 | 1.2027860 | 1.2027860 | 1.2027860 |
| $-\theta'(0)$ | 0.3 | 1.1847094 | 1.1467282 | 1.1463211 | 1.1463211 | 1.1463211 |
| | 0.7 | 1.0731585 | 1.0350043 | 1.0341356 | 1.0341344 | 1.0341344 |
| | 1.5 | 0.8462011 | 0.8114925 | 0.8093554 | 0.8093376 | 0.8093375 |
| | 0.1 | 1.2343382 | 1.2276842 | 1.2276800 | 1.2276800 | 1.2276800 |
| $-\phi'(0)$ | 0.3 | 1.2988048 | 1.2865458 | 1.2864829 | 1.2864829 | 1.2864829 |
| | 0.7 | 1.3245499 | 1.3072132 | 1.3069707 | 1.3069708 | 1.3069708 |
| | 1.5 | 1.3489087 | 1.3224155 | 1.3214809 | 1.3214790 | 1.3214790 |

**Table 7.** The effect of $D_f$ on skin-friction, heat and mass transfer coefficients when
$M = 0.5$, $Gr_x/Re^2 = 1.5$, $Gb = 0.5$, $R_d = 0.5$, $R = 2$ and $N_1 = 0.1$

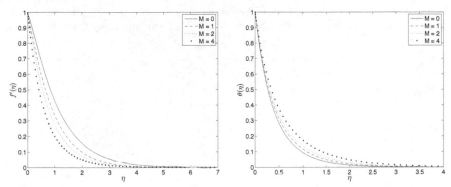

**Figure 1.** Effect of magnetic parameter $M$ on the (a) velocity and, (b) temperature when $Gr_x/Re^2 = 1.5$, $Gb = 0.5$, $R_d = 1$, $D_f = 0.3$, $Sr = 0.2$, $R = 0.1$ and $N_1 = 0.1$

The effects of the various fluid and physical parameters on the fluid properties are displayed qualitatively in Figures 1 - 7. Figure 1 illustrates the effect of the magnetic parameter $M$ on the boundary layer velocity and the temperature within the thermal boundary layer. As expected, we observe that increasing the magnetic filed parameter reduces the velocity due to an increase in the Lorentz force which acts against the flow if the magnetic field is applied in the normal direction. This naturally leads to an increase in the temperature (and concentration) within the boundary layer as less heat is conducted away.

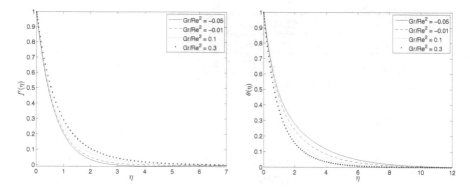

**Figure 2.** Effect of the mixed convection parameter $Gr_x/Re^2$ on the (a) velocity and (b) temperature when $M = 0.1$, $Gb = 0.1$, $R_d = 0.01$, $D_f = 0.3$, $Sr = 0.2$, $R = 0.2$ and $N_1 = 0.1$

Figure 2 shows the dimensionless velocity and temperature for various values of the mixed convection parameter $Gr_x/Re^2$ in the case of both aiding and opposing flow. We note that when the convection parameter increases, the velocity increases (the velocity is higher for aiding flow and less for opposing flow). The temperature (and solute concentration although not shown here) reduces as the convection parameter increases. Similar results were reported by Srinivasacharya and RamReddy [33].

Figure 3 shows the influence of the thermal radiation parameter $R_d$ and the viscous dissipation $Gb$ on the fluid velocity. The velocity increase with increasing thermal radiation parameter

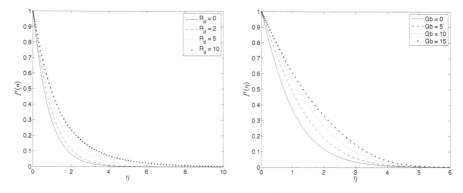

**Figure 3.** Effect of (a) the thermal radiation parameter $R_d$, and (b) viscous dissipation parameter $Gb$ on the fluid velocity when $M = 0.5$, $Gr_x/Re^2 = 1.5$, $Gb = 0.5$, $D_f = 0.3$, $Sr = 0.2$, $R = 0.1$ and $N_1 = 0.1$

Figure 4 shows the influence of the thermal radiation parameter $R_d$ and the viscous dissipation $Gb$ on the temperature within the thermal boundary layer. Naturally, the temperature increases with an increase in the thermal radiation and viscous dissipation parameters.

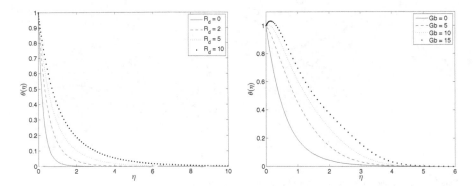

**Figure 4.** Effect of (a) the thermal radiation parameter $R_d$, and (b) viscous dissipation parameter $Gb$ on the temperature within the thermal boundary layer when $M = 0.5$, $Gr_x/Re^2 = 1.5$, $Gb = 0.5$, $D_f = 0.3$, $Sr = 0.2$, $R = 0.1$ and $N_1 = 0.1$

The effect of the viscous dissipation parameter $Gb$ on the solute concentration is shown in Figure 5. The solute concentration decreases with increasing viscous dissipation.

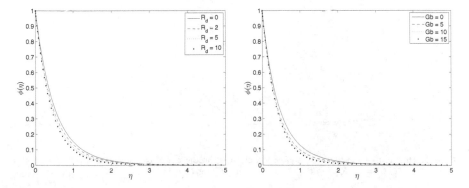

**Figure 5.** Effect of (a) the thermal radiation parameter $R_d$, and (b) the viscous dissipation parameter $Gb$ on the solute concentration when $M = 0.5$, $Gr_x/Re^2 = 1.5$, $R_d = 1$, $D_f = 0.3$, $Sr = 0.2$, $R = 0.1$ and $N_1 = 0.1$

In Figure 6 we show the effect of increasing the Dufour $D_f$ (that is, reducing the Soret $Sr$) parameter on the fluid velocity, temperature and solute concentration, respectively. The fluid velocity is found to increase with both parameters. An increase in $D_f$ enhances the temperature within the thermal boundary layer.

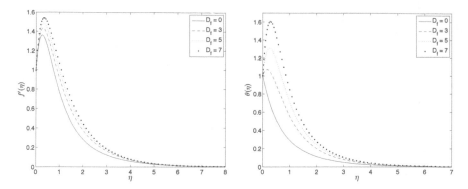

**Figure 6.** Effect of the Dufour number $D_f$ on the (a) velocity, and (b) temperature when $M = 0.5$, $Gr_x/Re^2 = 1.5$, $Gb = 0.5$, $R_d = 0.2$, $R = 0.1$ and $N_1 = 2$

The effect of the chemical reaction parameter $R$ on the fluid properties is shown in Figure 7. We note that the velocity reduces as the chemical reaction parameter $R$ increases. However, the solute concentration within boundary layer naturally decreases with an increase in the chemical reaction.

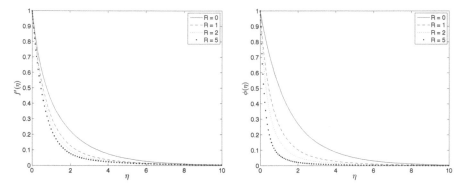

**Figure 7.** Effect of the chemical reaction parameter $R$ on the (a) velocity, and (b) concentration distributions when $M = 2$, $Gr_x/Re^2 = 1.5$, $Gb = 0.5$, $R_d = 1$, $R = 0.1$, $D_f = 0.3$, $Sr = 0.2$ and $N_1 = 5$

## 5. Conclusion

In this chapter we have studied the effects of cross-diffusion and viscous dissipation on heat and mass transfer from an exponentially stretching surface in porous media. We further considered the effects of thermal radiation and a chemical reaction. The governing equations were solved using the successive linearisation method. This has been shown to give accurate results. The effects of various physical parameters on the fluid properties, the skin-friction coefficient and the heat and the mass transfer rates have been determined. It was found, inter alia, that the velocity increase with the mixed convection parameter while the temperature and concentration profiles decrease. An increase in both viscous dissipation and radiation parameters reduced the concentration distribution while the temperature was enhanced by viscous dissipation and radiation parameters. The skin-friction, heat and mass transfer coefficients decreased with an increase in the magnetic field strength. The skin-friction and mass transfer coefficients decreased whereas the heat transfer coefficient increased with increasing Soret numbers.

## Author details

Ahmed A. Khidir and Precious Sibanda

School of Mathematics, Statistics and Computer Science, University of KwaZulu-Natal, South Africa

## References

[1] Al-Odat, M.Q.; Damseh, R.A. & Al-Azab, T.A. (2006). Thermal boundary layer on an exponentially stretching continuous surface in the presence of magnetic field effect. *Int. J. Appl. Mech. Eng.* 11, pp. 289-299.

[2] Ali, M. E. (1995). On thermal boundary layer on a power law stretched surface with suction or injection. *International Journal of Heat and Fluid Flow*, 16, pp. 280-290.

[3] Awad, F. G.; Sibanda, P.; Motsa, S. S. & Makinde, O. D. (2011). Convection from an inverted cone in a porous medium with cross-diffusion effects. *Computers & Mathematics with Applications*, 61, pp. 1431-1441.

[4] Awad, F. G.; Sibanda, P.; Narayana, M. & Motsa, S. S. (2011). Convection from a semi-finite plate in a fluid saturated porous medium with cross-diffusion and radiative heat transfer. *Int. J Physical Sciences*, 6, pp. 4910-4923.

[5] Awad, F.G.; Sibanda, P. & Narayana, M. (2011). *Heat and mass transfer from an inverted cone in a porous medium with cross-diffusion effects*. Mass Transfer-Advanced Aspects. pp. 81-106. InTech Open Access Publisher, Croatia.

[6] Bidin, B. & Nazar, R. (2009). Numerical solution of the boundary layer flow over an exponentially stretching sheet with thermal radiation. *Euro J. Sci. Res*, 33(4), pp. 710-717.

[7] Canuto, C.; Hussaini, M. Y.; Quarteroni, A. & Zang, T. A. (1988). *Spectral Methods in Fluid Dy- namics*. Springer-Verlag, Berlin.

[8] Don, W. S. & Solomonoff, A. (1995). Accuracy and speed in computing the Chebyshev collocation derivative. *SIAM J. Sci. Comput*, 16, pp. 1253-1268.

[9] Dulal, P. (2010). Mixed convection heat transfer in the boundary layers on an exponentially stretching surface with magnetic field. *App. Math. and Com*, 217, pp. 2356-2369.

[10] Dursunkaya, Z. & Worek, W. M. (1992). Diffusion-thermo and thermal diffusion effects in transient and steady natural convection from a vertical surface. *Int. J. Heat Mass Transfer*, 35, pp. 2060 -2065.

[11] Eckeret, E. R. G. & Drake, R. M. (1972). *Analysis of heat and mass transfer*. McGraw-Hill, New York.

[12] El-Aziz, M. A. (2008). Thermal-diffusion and diffusion-thermo effects on combined heat mass transfer by hydromagnetic three-dimensional free convection over a permeable stretching surface with radiation. *Physics Letter A*, 372, pp. 263-272.

[13] Elbashbeshy, E. M. A. (2001). Heat transfer over an exponentially stretching continuous surface with suction. *Archive of Mechanics*, 53, pp. 643-651.

[14] Ibrahim, S. Y. & Makinde, O. D. (2010). Chemically reacting MHD boundary layer flow of heat and mass transfer past a moving vertical plate with suction. *Scientific Research and Essay*, 5, pp. 2875-2882.

[15] Ishak, A. (2011). MHD boundary layer flow due to an exponentially stretching sheet with radiation effect. *Sains Malaysiana*, 40, pp. 391-395.

[16] Javed, T.; Sajid, M.; Abbas, Z. & Ali, N. (2011). Non-similar solution for rotating flow over an exponentially stretching surface. *Int. J. Numerical Methods for Heat and Fluid Flow*, 21, pp. 903-908.

[17] Khan, S. K. (2006). Boundary layer viscoelastic fluid flow over an exponentially stretching sheet. *International Journal of Applied Mechanics and Engineering*, 11, pp. 321-335.

[18] Loganathan, P.; Iranian, D. & Ganesan, P. (2011). Effects of chemical reaction on unsteady free convection and mass transfer flow past a vertical plate with variable viscosity and thermal conductivity. *European J. of Sc. Research*, 59, pp. 403-416.

[19] Magyari, E. & Keller, B. (1999). Heat and mass transfer in the boundary layers on an exponentially stretching continuous surface. *J Phys D: Appl Phy*, 32, pp. 577-585.

[20] Makukula, Z. G.; Sibanda, P. & Motsa, S. S. (2010). A note on the solution of the Von Kármán equations using series and chebyshev spectral methods. *Boundary Value Problems*, ID 471793 (2010).

[21] Makukula, Z.G.; Sibanda, P. & Motsa, S. S. (2010). On new solutions for heat transfer in a visco-elastic fluid between parallel plates. *Internationl Journal of Mathematical Models and Method in Applied Sciences*, 4, pp. 221-230.

[22] Makukula, Z.G.; Sibanda, P. & Motsa, S. S. (2010). A novel numerical technique for two-dimensional laminar flow between two moving porous walls. *Mathematical Problems in Engineering*, 2010, Article ID 528956. doi:10.1155/2010/528956.

[23] Motsa, S. S.; Sibanda, P. & Shateyi, S. (2011). On a new quasi-linearization method for systems of nonlinear boundary value problems. *Mathematical Methods in the Applied Sciences*, 34, pp. 1406-1413.

[24] Partha, M. K.; Murthy, P. V. S. N. & Rajasekhar, G. P. (2005). Effect of viscous dissipation on the mixed convection heat transfer from an exponentially stretching surface. *Heat Mass Transfer*, 41, pp. 360-366.

[25] Partha, M. K.; Murthy, P. V. S. N. & Rajasekhar, G. P. (2006). Soret and Dufour effects in a non-Darcy porous medium. *J. Heat Transfer*, 128, pp. 605-610.

[26] Raptis, A. (1998). Radiation and free convection flow through a porous medium. *International Communications in Heat and Mass Transfer*, 25, pp. 289-295.

[27] Sparrow, E.M. & Cess, R.D. (1978). *Radiation Heat Transfer*. Hemisphere, Washington.

[28] Sajid, M. & Hayat, T. (2009). Influence of thermal radiation on the boundary layer flow due to an exponentially stretching sheet. *Int. Comm. Heat Mass Transfer*, 35, pp. 347-356.

[29] Sanjayanand, E. & Khan, S. K. (2006). On heat and mass transfer in a viscoelastic boundary layer flow over an exponentially stretching sheet. *International Journal of Thermal Sciences*, 45, pp. 819-828.

[30] Sakiadis, B.C. (1961). Boundary layer behavior on continuous solid surfaces I: Boundary layer equations for two dimensional and axi-symmetric flow. *AIChE J*, 7, pp. 26-28.

[31] Sakiadis, B.C. (1976). Boundary layer behavior on continuous solid surfaces II. The boundary layer on a continuous flat surfaces. *AIChE J*, 7, pp. 221-225.

[32] Shateyi, S. & Motsa, S.S. (2010). Variable viscosity on magnetohydrodynamic fluid flow and heat transfer over an unsteady stretching surface with hall effect. *Boundary Value Problems*, 2010, Article ID 257568. doi:10.1155/2010/257568.

[33] Srinivasacharya, D. & RamReddy, Ch. (2011). Soret and Dufour Effects on Mixed Convection from an Exponentially Stretching Surface. *Int. J of Nonlinear Science*, 12, pp. 60-68.

# Chapter 6

# Modeling of Heat and Mass Transfer and Absorption-Condensation Dust and Gas Cleaning in Jet Scrubbers

M. I. Shilyaev and E. M. Khromova

Additional information is available at the end of the chapter

## 1. Introduction

The process of complex cleaning of gases, injected into the atmosphere, for instance, by thermal power plants, metallurgical, chemical or other industrial enterprises, from dust and harmful gaseous admixtures by means of their irrigation by wash liquids (water or specially selected water solutions) is considered. This process can be implemented in gas pipes or gas-cleaning apparatuses (direct flow or counter flow jet scrubbers) [1]. The process of gas cleaning from dust and gas admixtures is carried out in the following manner. The fluid dispersed by jets is introduced into the dust-vapor-gas flow in the form of droplets, interacts with it, and under nonisothermal conditions the increased moisture content leads to intensive condensation of liquid vapors on particles, their significant enlargement and efficient absorption of liquid droplets due to collisions of the latter with particles [2]. Simultaneously, the liquid droplets and condensate on particles absorb harmful gas components, dissolving them and removing from the vapor-gas flow.

The authors failed to find the mathematical description of this complex process in literature. From the engineering point of view the importance of development of generalized mathematical models, which reflect properly the interaction of heat and mass transfer with the effects of gas components removal and dust capture by the droplets of irrigating liquid in jet scrubbers and reactors, is undisputable, and it is determined by significant opportunities for optimization of operation conditions and constructions of energy-intensive and large-scale equipment in various industries both in terms of reducing of material and energy costs.

## 2. Problem statement, main equations and assumptions

In the current work we suggest the model for mathematical description of the above process
with the following assumptions:

1.  droplets and particles are considered monodispersed with equivalent sizes, equal to
    mass-median by distributions;

2.  concentrations of droplets of irrigating liquid, dust particles and harmful gas compo-
    nents are low, what allows us to use the Henry's law for equilibrium of gas components
    in liquid and gas phases at the interface and assume that the solution in droplet is ideal;

3.  the mean-mass temperature of droplets and temperature of their surfaces are equal be-
    cause of their small sizes [3];

4.  the typical time of gas component dissolution in droplet is significantly less than the
    typical time of mass transfer processes, commonly occurring in the apparatus;

5.  the motion velocities of particles with condensate on their surface ("formations") and
    vapor-gas flow are equal;

6.  the moisture content in the flow can be high, what requires consideration of the Stefan
    correction in mass transfer equations for evaporation-condensation process on droplets
    and "formations";

7.  we do not take into account the evaporation-condensation correction for the resistance
    and heat transfer coefficients of droplets and "formations", it is insignificant and be-
    comes obvious only at the initial stages of the process at high moisture contents [4];

8.  in equation of droplet motion we take into account variability of its mass;

9.  the radiant component in the process of heat transfer is neglected because of low tem-
    peratures of droplet, "formations" and flow;

10. mutual coalescence of droplets and "formations" is not taken into account, and merging
    of droplets and "formations" due to collision is the basis of condensation-inertial mech-
    anism of dust capture in jet scrubbers [2].

Under the above conditions equations of model system will take the following form:

Motion equation of a mass-median droplet with variable mass

$$\frac{d\vec{V}_d}{d\tau} = \vec{R}_d + \vec{g} - \frac{\vec{V}_d}{m_d} \frac{dm_d}{d\tau};$$  (1)

equation of heat transfer between droplet and vapor-gas flow

$$c_f m_d \frac{dT_d}{d\tau} = -\alpha_d \pi \delta_d^2 (T_d - T) + \sum r_i \frac{dm_{id}}{d\tau} + c_\delta \rho_\delta V_c \frac{\pi \delta_d^2}{4} \eta_{Stk}(T_\delta - T_0); \qquad (2)$$

equation of mass transfer between droplet and the i-$^{th}$ component of vapor-gas flow

$$\frac{dm_{i\delta}}{d\tau} = -\beta_{id} \pi \delta_d^2 \left(\rho_{id} - \rho_i\right); \qquad (3)$$

equation of mass transfer between "formation" and the i-$^{th}$ component of vapor-gas flow

$$\frac{dm_{i\delta}}{d\tau} = -\beta_{i\delta} \pi \delta^2 \left(\rho_{i\delta} - \rho_i\right); \qquad (4)$$

continuity equation for i-$^{th}$ reacting components, including vapor of liquid

$$\frac{\partial \rho_i}{\partial \tau} + div\left(\rho_i \vec{U}\right) = -\frac{dm_{id}}{d\tau} n_d - \frac{dm_{i\delta}}{d\tau} n_\delta; \qquad (5)$$

continuity equation for (mass concentration) of non-reacting component of the vapor-gas mixture

$$\frac{\partial \rho_g}{\partial \tau} + div\left(\rho_g \vec{U}\right) = 0; \qquad (6)$$

continuity equation for (mass concentration) of "formations"

$$\frac{\partial \rho_\delta}{\partial \tau} + div\left(\rho_\delta \vec{U}\right) = \sum \frac{dm_{i\delta}}{d\tau} n_\delta - \rho_\delta V_c \frac{\pi \delta_d^2}{4} \eta_{Stk} n_d; \qquad (7)$$

continuity equation for (mass concentration) of droplets

$$\frac{\partial \rho_d}{\partial \tau} + div\left(\rho_d \vec{V}_d\right) = \frac{dm_d}{d\tau} n_d; \qquad (8)$$

equation of heat transfer between "formation" and vapor-gas flow

$$\tilde{n}_\delta m_\delta \frac{\mathrm{d}T_\delta}{\mathrm{d}\tau} = -\alpha_\delta \pi \delta^2 \left(T_\delta - T\right) + \sum r_i \frac{\mathrm{d}m_{i\delta}}{\mathrm{d}\tau}; \tag{9}$$

equation of convective heat transfer between vapor-gas flow and droplets and "formations"

$$\rho \frac{\mathrm{d}\tilde{n}\left(T - T_0\right)}{\mathrm{d}\tau} = \alpha_d \pi \delta_d^2 \left(T_d - T\right) n_d + \alpha_\delta \pi \delta^2 \left(T_\delta - T\right) n_\delta; \tag{10}$$

general rate of droplet mass change due to evaporation-condensation and absorption of removed gas components (droplet collision is assumed unlikely) and "formation" absorption

$$\frac{\mathrm{d}m_d}{\mathrm{d}\tau} = \sum \frac{\mathrm{d}m_{id}}{\mathrm{d}\tau} + \rho_\delta V_c \frac{\pi \delta_d^2}{4} \eta_{\mathrm{Stk}} \tag{11}$$

general rate of "formation" mass change ("formation" collision is assumed unlikely )

$$\frac{\mathrm{d}m_\delta}{\mathrm{d}\tau} = \sum \frac{\mathrm{d}m_{i\delta}}{\mathrm{d}\tau}; \tag{12}$$

continuity equation for (mass concentration) of dry particles

$$\frac{\partial \rho_p}{\partial \tau} + div\left(\rho_p \vec{U}\right) = -\rho_p V_c \frac{\pi \delta_d^2}{4} \eta_{\mathrm{Stk}} n_d. \tag{13}$$

The following closure relationships shall be added to equations (1-13):

for the force of droplet aerodynamic resistance per a unit of droplet mass,

$$\vec{R}_d = -\xi \frac{\left(\vec{V}_d - \vec{U}\right)}{\tau_d}; \tag{14}$$

where relative coefficient of droplet resistance is $\xi = \xi / \xi_c$, $\xi_c = 24 / \mathrm{Re}_d$,

$$\tilde{\xi} = 1 + 0,197\,\mathrm{Re}_d^{0,63} + 2,6 \cdot 10^{-4}\,\mathrm{Re}_d^{1,38} \; (0,1 \le \mathrm{Re}_d \le 3 \cdot 10^5) \, \big[5\big], \tag{15}$$

$$\tau_d = \frac{\rho_f \delta_d^2}{18\mu}, \mathrm{Re}_d = \frac{V_c \delta_d \rho}{\mu}, V_c = \left|\vec{V}_d - \vec{U}\right|; \tag{16}$$

coefficient of "formation" entrainment according to the empirical formula of Langmuir–Blodgett with Fuchs correction on engagement effect [1]

$$\eta_{Stk} = \left(\frac{Stk}{Stk + 0.5}\right)^2 + 2.5\frac{\delta}{\delta_d}, \tag{17}$$

$$Stk = \tau_\delta \frac{V_c}{\delta_d}, \tau_\delta = \frac{\rho_{f\delta}\delta^2}{18\mu}\left(\rho_{f\delta} \approx \rho_f\right), \tag{18}$$

where $\rho_{f\delta}$ is efficient density of "formation';

mass transfer coefficient of the i-th component with droplets both via evaporation-condensation and absorption-desorption [2]

$$Nu_{id} = \frac{\beta_{id}\delta_d}{D_i} = 2\left(1 + 0,276Re_d^{0,5}Sc_i^{0,33}\right)K_{ci}, \; Sc_i = \frac{\mu}{\rho D_i}; \tag{19}$$

Stefan correction on increased moisture content

$$K_{\tilde{n}} = 1 + \frac{P_{id} + P_i}{2B}; \tag{20}$$

barometric (total) pressure

$$B = P_g + \sum P_i; \tag{21}$$

density of vapor-gas mixture

$$\rho = \rho_g + \sum \rho_i; \tag{22}$$

state equation for gas components and vapor of liquid

$$\rho_g = \frac{M_g P_g}{RT}, \; \rho_i = \frac{M_i P_i}{RT}, \; \rho_{id} = \frac{M_i P_{id}}{RT_d}, \; \rho_{i\delta} = \frac{M_i P_{i\delta}}{RT_\delta}; \tag{23}$$

diffusion coefficient of the i-th component in non-reacting component of the vapor-gas flow (we assume that its fraction in the flow is predominant)

$$D_i = D_{i0} \frac{B_0}{B}\left(\frac{T}{T_0}\right)^{1,75}, \quad B_0 = 0.1 \, MPa, \; T_0 = 273 \; K; \tag{24}$$

coefficient of droplet heat transfer according to Drake's formula

$$\mathrm{Nu}_d = \frac{\alpha_d \delta_d}{\lambda} = 2 + 0,459 \, \mathrm{Re}_d^{0,5} \, \mathrm{Pr}^{0,3}, \mathrm{Pr} = \frac{\mu c}{\lambda}; \tag{25}$$

countable concentrations of droplets and "formations"

$$n_d = \frac{\rho_d}{m_d}, \tag{26}$$

$$n_\delta = \frac{\rho_\delta}{m_\delta}; \tag{27}$$

heat and mass transfer coefficients of "formations"

$$\alpha_\delta = 2\frac{\lambda}{\delta} \;\; (\mathrm{Nu}_\delta = 2), \;\; \beta_{i\delta} = 2\frac{D_i}{\delta} \;\; (\mathrm{Nu}'_{i\delta} = 2), \tag{28}$$

heat capacity of the vapor-gas mixture

$$c = \frac{\sum \rho_i c_i}{\rho}; \tag{29}$$

specific heat of gas absorption with the made assumptions [6]

$$r_i = M_i^{-1} R T^2 \frac{\mathrm{d} \ln m_{px,i}}{\mathrm{d}T}, \tag{30}$$

it can be assumed for water vapors that $r_d \approx 2500$ kJ/kg [2-4];

according to Henry's law for partial saturation pressure at the interface between i-[th] gas components, the equilibrium condition is [6]

$$P_{id} = m_{px,i} x_{id}, \quad P_{i\delta} = m_{px,i} x_{i\delta}; \tag{31}$$

$$x_{id,\delta} = \frac{\dfrac{c_{mid,\delta}}{M_i}}{\dfrac{c_{mid,\delta}}{M_i} + \dfrac{1}{M_{dis}}}, \tag{32}$$

where $x_{id,\delta}$ is a molar part, equal to the number of moles of dissolved gas per the total number of moles in solution, $M_{dis}$ is the molar mass of dissolvent.

The equation for mass concentration of dissolved i-[th] gas component in the kilogram per 1 kg of dissolvent in the droplet and "formation" is written as

$$\frac{dc_{mid,\delta}}{d\tau} = \frac{dm_{id,\delta}}{d\tau} \frac{6}{\pi \delta_{d,f}^3 \rho_f}. \tag{33}$$

Diameters of specific spherical volume of dissolvent for "formation" $\delta_f$ and droplet $\delta_d$ are calculated by equations:

$$\frac{d\delta_f^3}{d\tau} = \frac{6}{\pi} \frac{1}{\rho_f} \frac{dm_{v\delta}}{d\tau}, \delta_d = \sqrt[3]{\frac{6m_d}{\pi \rho_f}}; \tag{34}$$

"formation" diameter is

$$\delta = \sqrt[3]{\frac{6m_\delta}{\pi \rho_s} + \delta_0^3}, \left(\rho_s \approx \rho_f\right), \tag{35}$$

where $\varrho_s$ is solution density, kg/m$^3$.

The reactive force in equation (1) is neglected because of evaporation-condensation and absorption [2]. In equation (2) specific heat capacity $c_f$ is taken constant and equal to specific heat capacity of dissolvent because of low concentrations of absorbed dust and absorbed gases. For small particles and significant amount of condensate on them [2] we will take $c_\delta$ equal to specific heat capacity of dissolvent $c_f$. In this equation the first summand in the right determines convective heat transfer between the droplet and flow, the second summand determines the total heat of phase transitions due to evaporation-condensation and absorption-desorption of gas components, and the third summand determines heat introduced by "formations" into the droplet due to their absorption at collision. In equations (3) and (4) $dm_{id}/d\tau$, $dm_{i\delta}/d\tau$ are the rates of droplet and "formation" mass change due to the processes of evaporation-condensation or absorption-desorption of the i-[th] gas component. In relationships (16) dynamic viscosity of the vapor-gas flow $\mu$ is calculated by generalized Wilkey's formulas, in our case on the basis of research performed in [2] and [4] with consideration of low concentrations of reacting gas components we will determine $\mu$ by Sutherland formulas [2, 4] for a non-reacting component of the vapor-gas mixture. The coefficient of mixture heat conductivity $\lambda$ will be calculated similarly by Sutherland formula [2, 4] in formulas (25) and (28). The diffusion components of the vapor-gas flow will be determined by their dependences on temperature in the non-reacting component by formula (24). The correction for Stefan flow of gas components is not taken into account because of their low concentrations ($K_{ci}$=1). In the current study we will consider water as the absorbent and $m_{px,i}$ will be taken from tables depending on temperature [6]. If there are no data for some gases in [6], for instance, for $SO_2$, we suggest to recalculate volumetric 1 and weight $q_s$ solubility [7] as the limit ones by $m_{px,i}$, this will be described in detail in this work.

## 3. Numerical implementation of the model, comparison of calculation results with experimental data

As it is shown in [2], in most technically implemented situations it is possible to use a single-dimensional model for calculation of heat and mass transfer in irrigation chambers, what is determined by the vertical position of apparatuses (hollow jet scrubbers HJC); at their horizontal position it is determined by high velocities of cleaned gases, dust particles and droplets (Venturi scrubber VS), when the gravity force, influencing the flow components and causing its 2D character, is low in comparison with the inertia forces.

The calculation scheme of the problem for the vertical construction of apparatus is shown in Fig. 1a). The scheme of interaction between a droplet of washing liquid dispersed by the jets with vapor-gas flow and dust particles is shown in Fig. 1b).

The hollow jet scrubber HJS can have direct-flow and counter-flow construction. In the direct-flow scheme the initial parameters of the vapor-gas flow, irrigating liquid and dust are set on one side (inlet) of apparatus, and the resulting parameters are achieved at the apparatus outlet. In the counter-flow scheme the parameters of vapor-gas flow and dust are set on one side of apparatus, the parameters of irrigating liquid are set on the opposite side (at apparatus outlet). Scheme 1a) is attributed to the counter-flow. From the point of numerical implementation the direct-flow scheme is the Cauchy problem, and the counter-flow scheme is the boundary problem. Let's perform calculations for the direct-flow scheme according to the known experimental data for generalized volumetric mass transfer coefficients, shown in [6, p. 562], for different gases absorbed on dispersed water. The calculation scheme is shown in Fig. 2 (it is conditional, the construction can differ).

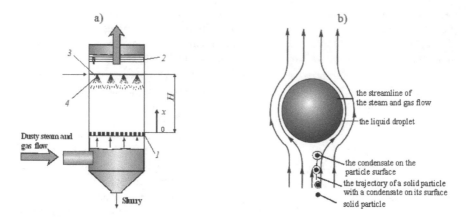

**Figure 1.** The HJS scheme: *1* – gas-distributing grate, *2* – droplet catcher, *3* – water collector, *4* – jets, *H* – scrubber operation height; b) the scheme of droplet interaction with the flow and dust particle

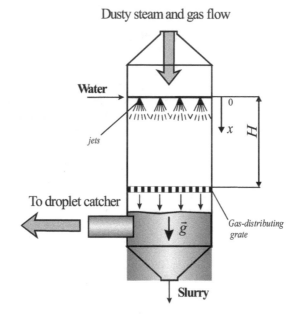

**Figure 2.** The scheme of direct-flow HJS

The problem will be solved in the stationary statement. The boundary conditions are set at $x=0$ ($\tau=0$) in the following manner:

for the vapor-gas flow $\quad U = U_0, \ d = d_0, \ d_i = d_{i0}, \ T = T_{00};$

for dispersed liquid $\qquad V_d = V_{d0}, \ \delta_d = \delta_{d0}, \ q = q_0, \ T_d = T_{d0};$

for dust $\qquad\qquad\qquad \rho_p = \rho_{p0}, \ \delta = \delta_0.$

$$(36)$$

Continuity equations (6) and (8) in stationary single-dimensional case can be reduced to the following, as in [2], analytical dependences:

$$U = U_0 \frac{T}{T_{00}} \frac{B - \sum P_{j0}}{B - \sum P_j} \approx U_0 \frac{T}{T_{00}} \frac{K_{ave} + d_\Sigma}{K_{ave} + d_{\Sigma 0}}; \tag{37}$$

$$\rho_d = \rho_{d0} \frac{V_{d0}}{V_{dx}} \frac{m_d}{m_{d0}}, \ \rho_{d0} = q \rho_f \frac{U_0}{V_{d0}}, m_d = \rho_f \frac{\pi \delta_d^3}{6}, m_{d0} = \rho_f \frac{\pi \delta_{d0}^3}{6}, \tag{38}$$

where

$$K_{ave} = \frac{\sum_1^k K_i}{k}, K_i = \frac{M_i}{M_g}, \tag{39}$$

$k$ is the number of reacting components, including liquid vapors, $M_g$ is molecular mass of a non-reacting component of gas;

$$d_\Sigma = \frac{\sum_1^k K_i P_i}{B - \sum_1^k P_i} \approx K_{ave} \frac{\sum_1^k P_i}{B - \sum_1^k P_i}, \tag{40}$$

efficiency of dust capture and gas component removal is determined by relationships:

$$\eta_p = 1 - \frac{\left(\rho_p U\right)_H}{\rho_{p0} U_0}, \tag{41}$$

$$\eta_i = 1 - \frac{\left(\rho_{d.a} d_i U\right)_H}{\rho_{d.a0} d_{i0} U_0}. \tag{42}$$

## 3.1. Calculation with the use of $m_{px}$ [6] or Henry's coefficient E [7]

Calculation results on absorption of $CO_2$ by water droplets in the direct-low hollow jet scrubber are shown in Fig. 3. There are no any restrictions for calculations by solubility limit (concentration of gas dissolving in the droplet). However, the solubility limits exists as the experimental fact for gases, presented in tables of Hand-books as absorption coefficient $\alpha$, in volumetric fractions reduced to 0 °C and pressure of 0.1 MPa or in the form of solubility co-efficient $q_p$ in mass fractions to solution or dissolvent [8, 9]. Thus, in [9] for $\alpha(T)$, $m^3$ of gas/ $m^3$ of water, for $CO_2$ and $CH_4$ the following data are shown (Table 1).

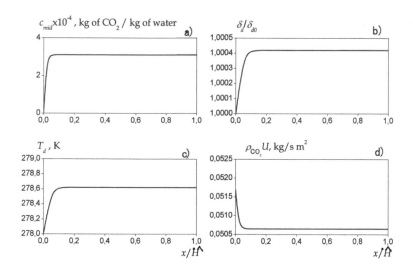

**Figure 3.** Calculation of $CO_2$ absorption in direct-flow jet scrubber: $H$ =12.75 m; $q$ =0.015 $m^3/m^3$; $\delta_{d0}$ =700 μm; $V_{d0}$ =24.5 m/s; $U_0$ =0.25 m/s; $T_{d0}$ =278 K; $T_{00}$ =293 K; $d_0$ =0.02 kg/kg of dry air; $d_{CO_2}$ =0.2 kg of $CO_2$/kg of dry air; $\eta_{CO_2}$ =0.022262

| $T, K$ | 273 | 283 | 293 | 303 | 313 | 323 | 333 | 353 | 373 |
|---|---|---|---|---|---|---|---|---|---|
| $\alpha_{CO_2}$ | 1.713 | 1.194 | 0.878 | 0.665 | 0.530 | 0.436 | 0.359 | ... | ... |
| $\alpha_{CH_4} \bullet 10^3$ | 55.6 | 41.8 | 33.6 | 27.6 | 23.7 | 21.3 | 19.5 | 17.7 | 17.0 |

**Table 1.** Absorption coefficients for $CO_2$ and $CH_4$ $a$, $m^3/m^3$ of water at $B$=0.1 MPa and $T_0$=273 K

It follows from this table that the limit value of $CO_2$ concentration in a water droplet is

$$c_{mco_2,\lim} = \alpha_{co_2}\rho_{co_2}(T_0, B_0) = \alpha_{co_2}(T)\frac{M_{co_2}B_0}{RT_0} = \alpha_{co_2}\frac{44 \cdot 101325}{8,314 \cdot 10^3 \cdot 273} =$$
$$= 1,964 \cdot 10^{-3} \cdot \alpha_{co_2}(T), \quad \frac{\kappa g CO_2}{\kappa g \, water}, \tag{43}$$

where $\alpha_{co_2}(T)$ is the table value of absorption coefficient for $CO_2$ (Table 1).

This value shall limit concentration of $CO_2$ dissolved in the droplet. It can be seen in Fig. 3a) that according to Table 1 calculated value of $c_{mid}$ does not reach the solubility limit and it is one order lower. Thus, $c_{mid,\lim}$ ($T_d$ =278, 63 K, see Fig. 3c)) =2.81•$10^{-3}$ kg of $CO_2$/kg of water. It can be seen in Fig. 3b) that as a result of water vapor condensation and $CO_2$ absorption the size of droplet increases insignificantly, less than by 0.05 %, i.e., a small amount of water vapors condenses on the droplet and a small amount of $CO_2$ is absorbed by the droplet, Fig. 3a). The calculated amount of the mass of gas component absorbed by liquid droplets in the scrubber is determined by formula, kg/h•m²,

$$W_{i,th} = \rho_{d.a.,0}d_{i0}U_0\eta_i 3600. \tag{44}$$

For calculated situation with consideration of partial density of dry air at the inlet $\rho_{d.a.,0}$ =1.0363 kg/m³, gas content $d_{CO_2,0}$=0.2 kg/kg of dry air, $U_0$=0.25 m/s and $\eta_{CO_2}$=0.022262 according to formula (44) we will obtain

$$W_{i,th} = 1,0363 \cdot 0,2 \cdot 0,25 \cdot 0,022262 \cdot 3600 = 4,153 \; kg/h \cdot m^2. \tag{45}$$

Let's compare the obtained result with the value achieved via the empirical volumetric mass transfer coefficient, shown in [6, p. 562] (in our nomenclature):

$$\beta_{iv} = 720U_0^{0,9}Q^{0,45}H^{-0,65}\left(\frac{D_{i,0}}{D_{NH_3,0}}\right)^{0,67}, \; 1/h. \tag{46}$$

Here $Q$ is irrigation density, m/h, $D_{NH_3,0}$=0,198 $10^{-4}$ m²/s is coefficient of methane diffusion in air at $B_0$=0.1 MPa and $T_0$=273 K, H is calculated scrubber height, m.

Let's write down the value of obtained coefficient per an area unit of apparatus cross-section via coefficient $\beta_{iv}$ in the following form, kg/h•m²:

$$W_{i,e} = \beta_{iv}\Delta\rho_i l, \tag{47}$$

where $\Delta\rho_i$ is calculated concentration pressure on the way of gas component obtaining $x=0$, $x=l$, where $l$ corresponds to the coordinate, where thermodynamic equilibrium is achieved for the i-th component in the flow.

Substituting (46) into (47) and assuming $l=H$, as it was made at treatment of experimental data in [6], we get

$$W_{i,e} = 720U_0^{0,9}\left(3600U_0q\right)^{0,45}H^{0,35}\Delta\rho_i\left(\frac{D_{i,0}}{D_{NH_3,0}}\right)^{0,67}, \text{kg/h}\cdot\text{m}^2, \qquad (48)$$

where q is irrigation coefficient, $m^3$ of water/ м$^3$ of vapor-gas flow at apparatus inlet.

Let's transform formula (48), and finally for calculation we obtain dependence

$$W_{i,e} = 28685,9U_0^{1,35}q^{0,45}H^{0,35}\Delta\rho_i\left(\frac{D_{i,0}}{D_{NH_3,0}}\right)^{0,67}. \qquad (49)$$

For the considered situation $D_{i,0}=D_{CO_2,0}=0,138\cdot10^{-4}m^2/s$ [6].

Let's take the average experimental data of [6] as the calculation working height of absorber $H_{ave}=(H_{min} \quad H_{max})^{1/2}=(4.3 \quad 12)^{1/2}\approx7$ m, then for calculated value $\Delta\rho_{CO_2}=\rho_{d.a.,0}d_{CO_2,0}\eta_i=1,0363\cdot0,2\cdot0,022262=0,004614$ kg/m$^3$

$$W_{i,e} = 28685,9\cdot(0,25)^{1,35}(0,015)^{0,45}(7)^{0,35}0,004614\left(\frac{0,138}{0,198}\right)^{0,67}\approx4,772 \text{ kg/h}\cdot\text{m}^2 \qquad (50)$$

since $D_i=D_{i,0}\dfrac{B_0}{B}\left(\dfrac{T}{T_0}\right)^{1,75}$ and total multiplier $\dfrac{B_0}{B}\left(\dfrac{T}{T_0}\right)^{1,75}$ in (50) is reduced.

Calculated (45) and experimental (50) results differ by $\Delta\approx13$ %. If we take $H=4.3$ m, then $W_{i,e}=4,024$ kg/h m$^2$ and $\Delta\approx3$ %. At $H=12$ m, $W_{i,e}=5,76$ kg/h m$^2$ and $\Delta=28$ %.

It follows from formula (42) for calculated concentration difference at apparatus inlet and outlet that

$$\Delta\rho_{i,cal} = \left(\frac{U_{out}}{U_0}-1\right)\rho_{iin}+\Delta\rho_i,$$

where calculated value is $\Delta\rho_{i,cal}=\eta_i\rho_{i,in}$. Thus, for $U_{out}=U_0$ $\Delta\rho_{i,cal}=\eta_i\rho_{i,in}=\Delta\rho_i$, i.e., the calculated value of concentration difference coincides with real value $\Delta\rho_i$. Therefore, we can make a conclusion that efficiency of dust capture and mass transfer shall be calculated not by the measured difference of dust concentrations and extracted gas components at the inlet and outlet, but by difference of their mass fluxes in accordance with the law of mass conservation, and if velocities at the inlet and outlet are equal or close efficiencies can be calculated by real concentration difference. Distribution of mass flux of $CO_2$ along the scrubber height is shown in Fig. 3d). It is obvious from Figs. 3c) and 3d) that the process of absorption completes long before the flow escape from the scrubber, for the given version of calculation at $x/H \approx 0.1$ ($\approx 1.3$ m). Hence, the residual height of the scrubber is excessive, and it can not be determined experimentally.

Previous comparison can be made in the relative form, what will prove the validity of calculation of mass transfer coefficient as a measure determining process intensity, on the basis of model in comparison with its experimental expression [6]:

$$\frac{W_{i,th}}{W_{i,e}} = \frac{\rho_{d.a.,0} d_{i0} U_0 \eta_i \, 3600}{28685,9 U_0^{1,35} q^{0,45} H^{0,35} \Delta \rho_i \left( \dfrac{D_{i,0}}{D_{NH_3,0}} \right)^{0,67}} = \frac{0,1255}{q^{0,45} (HU_0)^{0,35} \left( \dfrac{D_{i,0}}{D_{NH_3,0}} \right)^{0,67}}, \tag{51}$$

where it is assumed that $\Delta \rho_i = \eta_i \rho_{d.a.,0} d_{i0}$. Thus, for our case $\dfrac{W_{i,th}}{W_{i,e}} = \dfrac{0,1255}{0,616 \cdot 0,151 \cdot H^{0,35} \cdot 0,785} = \dfrac{1,72}{H^{0,35}} = 0,87$ at $H=7$ m and $\dfrac{W_{i,th}}{W_{i,e}} = 1,03$ at $H=4.3$ m.

Here $\Delta \rho_i$ has the meaning of efficient drop of gas concentration, not real, but corresponding to extraction of the gas component due to absorption on a liquid droplet. Real drop of $CO_2$ concentrations at the inlet and outlet at the example of Fig. 4 is even negative: $\Delta \rho_{CO_2} = \rho_{CO_2,in} - \rho_{CO_2,out} = 0,1245 - 0,2099 = -0,0854$ kg/m³, here $U_{in} = U_0 = 0.25$ m/s, $U_{out} = 0.1439$ m/s.

Extractions per a total volume of apparatus can be presented as, kg/h,

$$\Delta G_{i,th} = \beta_{iv,th} \Delta \rho_i \pi D^2 H; \tag{52}$$

$$\Delta G_{i,e} = \beta_{iv,e} \Delta \rho_i \pi D^2 H. \tag{53}$$

On the other hand

$$\beta_{iv,th} = \frac{\Delta G_{i,th}}{\Delta \rho_i \pi D^2 H} = \frac{\rho_{d.a.,i0} d_{i0} U_0 \eta_i 3600 \pi D^2}{\Delta \rho_i \pi D^2 H}.$$

Hence, with consideration of formulas (46) and (53) we will obtain the relationship for volumetric mass transfer coefficients (theoretical and experimental ones)

$$\frac{\beta_{iv,th}}{\beta_{iv,e}} = \frac{3600 U_0 \Delta\rho_i \pi D^2}{28685,\, 9U_0^{1,35} q^{0,45} H^{-0,65} \Delta\rho_i \left(\dfrac{D_{i,0}}{D_{NH_3,0}}\right)^{0,67} \pi D^2 H},$$

after elementary reductions in numerator and denominator this corresponds to formula (51). Here $D$ is apparatus diameter.

Calculations results for the same situation as in Fig. 3 are shown in Fig. 4, but for the increased moisture content $d_0$=0.5 kg/kg of dry air. The theoretical value of absorbed $CO_2$ is:

$$W_{i,th} = 0,\, 6226 \cdot 0,\, 2 \cdot 0,\, 25 \cdot 0,\, 029136 \cdot 3600 = 3,\, 265\ \text{kg/h} \bullet \text{m}^2.$$

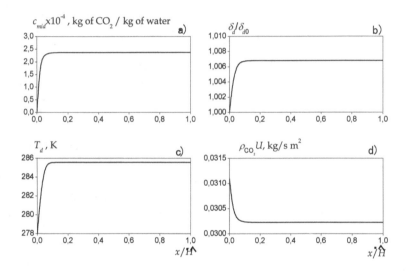

**Figure 4.** Fig. 4. Calculation of $CO_2$ absorption in the direct-flow jet scrubber: $H$ =12.75 m; $q$ =0.015 m³/m³; $\delta_{d0}$ =700 µm; $V_{d0}$ =24.5 m/s; $U_0$ =0.25 m/s; $T_{d0}$ =278 K; $T_{00}$ =293 K; $d_0$ =0.5 kg/kg of dry air; $d_{CO_2}$ =0.2 kg of $CO_2$/kg of dry air; $n_{CO_2}$ =0.029136

Calculation by formula (48) for height $H$=4.3 m gives the following

$$W_{i,e} = 28685,\, 9 \cdot (0,\, 25)^{1,35}(0,\, 15)^{0,45}(4,\, 3)^{0,35}0,\, 00363 \cdot 0,\, 785 = 3,\, 165\ \text{кg/h} \bullet \text{m}^2,$$

what differs from the theoretical value by 3 %. Here $\Delta\rho_{CO_2}$=0, 00363 kg/m³ by calculation ($\Delta\rho_{CO_2}$=0, 62226 · 0, 2 · 0, 029136). We should note that even for the increased moisture contents the size of droplets increases slightly due to condensation and absorption (less than by 1 %) (Fig. 4b). For calculated scrubber height $H_{ave}$=7 m $W_{i,e}$=3, 75 kg/h•m² (Δ=15.6 %).

According to comparison, the model agrees well with the experimental data.

In calculations tabular data $m_{px}$ for water solution of $CO_2$ [6] were approximated by temperature dependence $T$,

$$m_{px} = \left(2,389T^2 - 994,6T + 100765\right)10^4, \text{Pa}. \tag{54}$$

Partial pressures of saturated water vapors on droplet and "formation" surfaces were calculated by formula [?] (the partial pressure of saturated vapors of gas components were not taken into account)

$$P_{sd,\delta} = P_{cr}\exp\left(A_1\ln\frac{T_{d,\delta}}{T_{cr}} + A_2 f_2\right), \tag{55}$$

where

$$f_2 = \frac{4\left(\dfrac{T_{d,\delta}}{T_{cr}} - 1\right)}{\dfrac{T}{T_{cr}}} + f_1 - 5,3\ln\frac{T_{d,\delta}}{T_{cr}},$$

$$f_1 = \left(\frac{T_{d,\delta}}{T_{cr}} - 1\right)\left[\frac{\left(\dfrac{T_{d,d}}{T_{cr}} + 1\right)^2}{5} + 0,5\right],$$

$P_{cr} = 221.29 \ 10^5$ Па; $t_{cr} = 374.1$ °C; $A_1 = 7.5480$; $A_2 = 2.7870$.

For hydrogen sulfide $m_{px}$ for water solution [6] was approximated by dependence:

$$m_{px} = \left(-0,0251T^2 + 148,73T - 36374\right)10^4, \text{Pa}. \tag{56}$$

Calculation results on absorption of hydrogen sulfide on a water droplet from the vapor-gas flow are shown in Fig. 5.

Theoretical value of $W_{i,th}$ for $H_2S$ ($\rho_{d.a.,0} = 1,0029$kg/m³) is

$$W_{i,th} = 1,0029 \cdot 0,2 \cdot 0,25 \cdot 0,062478 \cdot 3600 = 11,28 \ \text{kg/h m}^2. \tag{57}$$

Calculation by formula (48) with experimental mass transfer coefficient gives for $H$=4.3 m

$$W_{i,e} = 28685,9 \cdot 0,1539 \cdot 0,151 \cdot 1,66615 \cdot 0,01253 \cdot 0,7426 = 10,335 \text{ kg/h m}^2, \tag{58}$$

where $\left(\dfrac{D_{H_2S}}{D_{NH_3}}\right)^{0,67} = \left(\dfrac{0,127}{0,198}\right)^{0,67} = 0,7426,$   $\Delta\rho_{H_2S} \approx \eta\rho_{SO_2,0} \approx 0,22031 \cdot 1,0029 \cdot 0,2 = 0,01253$

kg/m³. Difference between results of (57) and (58) is $\Delta\approx8$ %. For calculated height $H$=7 m $W_{i,e}$=12, 257 kg/h m² and $\Delta\approx8$ % on the other hand. In calculations for H₂S the limit of concentration (solubility) in water is not exceeded (solubility for 20 °C is about 3.85 10⁻³ kg of H₂S/kg of water) (see Fig. 5a)). According to the diagrams, here absorption is completed at 1.3 – 1.5 m from the scrubber inlet.

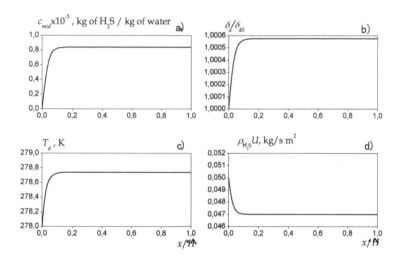

**Figure 5.** Calculation of hydrogen sulfide absorption: $H$ =12.75 m; $q$ =0.015 m³/m³; $\delta_{d0}$ =700 μm; $V_{d0}$ =24.5 m/s; $U_0$ =0.25 m/s; $T_{d0}$ =278 K; $T_{00}$ =293 K; $d_0$ =0.02 kg/kg of dry air; $d_{H_2S,0}$ =0.2 kg of H₂S/kg of dry air; $\eta_{H_2S}$=0.062478

### 3.2. Calculation of absorption by solubility of l and q$_s$

If there are no tabular data for $m_{px}$ (or $E$) of any gas, and solubility information is available in the hand-book, for instance, for $l$, m³ of gas/m³ of water and for $q_s$, g of gas/100 g of water, we can relate $l$ and $q_s$ to the limit density of saturated gas on the droplet surface $\rho_{id,lim}$, kg/m³, taking into account that the process of its dissolution occurs in droplet volume fast, i.e., the typical time of gas dissolution is significantly less than the typical time of droplet stay in the working volume of scrubber. Then,

$$\rho_{id,\lim} = 10\frac{q_s}{l}. \tag{59}$$

Thus, in [8, p. 260-261] there are tabular data for $SO_2$ for $l$ and $q_s$, where we have shown re-calculation of $\rho_{SO_2d,\lim}$ by formula (59) in the last line of Table 2:

| $t$, °C | 0 | 10 | 20 | 30 | 40 |
|---|---|---|---|---|---|
| $l$ | 79.8 | 56.7 | 39.4 | 27.2 | 18.8 |
| $q_s$ | 22.8 | 16.2 | 11.3 | 7.8 | 5.41 |
| $\rho_{SO_2d,\lim} = 10\frac{q_s}{l}$ | 2.8571 | 2.8571 | 2.8680 | 2.8676 | 2.8777 |

**Table 2.** Volumetric $l$ and weight $q_s$ solubility coefficients for SO

According to this Table, $\rho_{SO_2d,\lim}=2.8655\approx2.9$ kg/m³ and it is almost constant value.

First, for this case we calculate $m_{px}$:

$$P_{id,\lim} = P_{SO_2d,\lim} = \rho_{SO_2d,\lim}\frac{RT_d}{M_{SO_2}} = 2,8655\frac{8,314\cdot10^3}{64}T_d; \tag{60}$$

$$\left(m_{px}\right)_{SO_2} = \frac{P_{SO_2d,\lim}}{x_{SO_2d,\lim}}; \tag{61}$$

$$x_{SO_2d,\lim} = \frac{\dfrac{q_s10^{-2}}{M_{SO_2}}}{\dfrac{q_s10^{-2}}{M_{SO_2}} + \dfrac{1}{M_{water}}}. \tag{62}$$

As a result, the following approximation was obtained by formula (61) for $SO_2$

$$\left(m_{px}\right)_{SO_2} = 2976,58T^2 - 1594158T + 215090898, \quad \text{Pa.,} \tag{63}$$

Knowing $\left(m_{px}\right)_{SO_2}$, we determine specific heat of $SO_2$ absorption by water:

$$RTM_{SO_2}^{-1} \frac{d\ln\left(m_{px}\right)_{SO_2}}{dt} = r_{SO_2}\left(T_d\right), \text{ J/kg.} \tag{64}$$

Following calculation is performed by the general scheme (formulas (31)–(35)).

Results of calculation are shown in Fig. 6 at ventilation of air humidity $d_0$=0.02 kg/kg of dry air, $d_{SO_2,0}$=0.2 kg/kg of dry air, $d_{CH_4,0}$=0.2 kg/kg of dry air. Other parameters are shown in captions to the figure. It was obtained for this calculation version that $\eta_{SO_2}$=51.1 %, $\eta_{CH_4}$=0.08 %. The limit value of $SO_2$ concentration in a droplet is not achieved even for $CH_4$. According to tabular data on absorption coefficient $\alpha$, $m^3/m^3$ of water (see Table 1):

$c_{mid,\lim}$=0, 717 · $10^{-6}\alpha$, kg of $CH_4$/ kg of water.

According to calculation of extracted $SO_2$ for the given case ($\rho_{d.a.,0}$=0, 8121kg/m³):

$W_{i,th}$ =0, 8121 · 0, 2 · 900 · 0, 51072=74, 656  kg/h•m²,

$W_{i,e}$=28685, 9 · 0, 1539 · 0, 151 · (12)$^{0,35}$ · 0, 8121 · 0, 2 · 0, 51072 · 0, 715=94, 34  kg/h•m²

for $H$=12 m,

$W_{i,e}$=28685, 9 · 0, 1539 · 0, 151 · (7)$^{0,35}$ · 0, 8121 · 0, 2 · 0, 51072 · 0, 715=78, 13  kg/h•m²

at $H_{ave}$=7 m ($H_{ave}=\sqrt{4, 3 \cdot 12}\approx$7m). Here $\left(\dfrac{D_{SO_2,0}}{D_{NH_3,0}}\right)^{0,67}$=0, 715.

Comparison of $W_{i,th}$ and $W_{i,e}$ for $SO_2$ proves good agreement between theory and experiment.

According to calculation, Fig. 6c), methane is not absorbed by water. However, even for methane comparison of calculation with experiment yields satisfactory agreement:

$W_{i,th}$ =0, 1624 · 900 · 0, 00079541=0, 11626 kg/h•m²;

$W_{i,e}$=28685, 9 · 0, 1539 · 0, 151 · 0, 1624 · 0, 00079541 · 0, 715=0, 1026 kg/ h•m²

at $H$=4.3 m, $\Delta$=11.75 %. At $H$=7 m, $W_{i,e}$=0, 1217 kg/ h•m² and $\Delta$=4.5 %.

We should note that absorber height $H$ in experimental dependence for $\beta_{iv}$ is taken improperly. The optimal and calculated height of setup should equal path $l$, where the process of component extraction is completed. In most cases of calculations, it completed earlier at the height less than the accepted height of absorber $H$=12.75 m. Therefore, at comparison of calculation and experimental data in experimental dependence for mass transfer coefficient we have varied the calculated height in the range of the heights of tested setups from 4.3 to 12 m [6].

For $CH_4$ $m_{px}$ is approximated by dependence

$$m_{px} = \left(-47,\ 154T^2 + 35490T - 5962310\right) 10^4, \quad \text{Pa}. \tag{65}$$

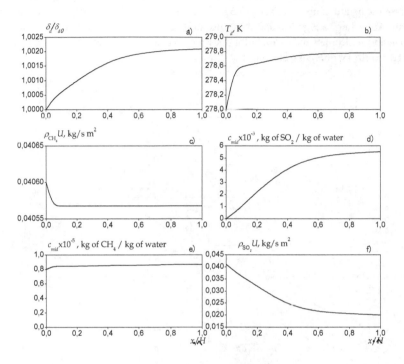

**Figure 6.** Calculation of $SO_2$ and $CH_4$ absorption in direct-flow jet scrubber: $H = 12.75$ m; $q = 0.015$ m³/m³; $\delta_{d0} = 700$ μm; $V_{d0} = 24.5$ m/s; $U_0 = 0.25$ m/s; $T_{d0} = 278$ K; $T_{00} = 293$ K; $d_0 = 0.02$ kg/kg of dry air; $d_{SO_2,0} = 0.2$ kg/kg of dry air; $d_{CH_4,0} = 0.2$ kg/kg of dry air; $\eta_{SO_2} = 0.51064$; $\eta_{CH_4} = 0.00079541$

## 3.3. Calculation of combined absorption-condensation dust-gas cleaning

Calculations of combined condensation dust capture and absorption extraction of hydrogen sulfide from the vapor-air flow in direct-flow hollow scrubber are shown in Fig. 7. Calculated parameters are shown below the figure. According to Fig. 7a), even at increased moisture content the size of droplets increases weak due to condensation. Therefore, for similar processes the equation of droplet motion can be calculated with a constant mass.

An increase in the size of "formations" is more significant due to condensation of water vapors on them: for $\delta_0 = 0.01$ μm it is 2.1, for $\delta_0 = 0.1$ μm it is 2.3, and for $\delta_0 = 1$ μ it is 32 and more for the same total concentration of dust at the inlet of 1.72 g/m³. In the first case, particles are not caught, in the second case, about 5.76 % of particles are caught, and in the third case, 100 % of particles are caught at the inlet to the apparatus. For this version of calculation the stable state by concentrations of $H_2S$ dissolved in droplets and in condensate on "formations" occurs far from the flow escape from the scrubber. For particles water vapor condensation at flow escape from the scrubber has been also competed already (see Fig. 7g). Therefore, in this case the height of absorber above 1.5 m is excessive, and in construction it can be limited by 2 m. According to Figs. 7d) and 7e), concentration of $H_2S$ dissolved in condensate on the particle and in droplets increases, but it does not exceed the solubility limit (in this case it is about $3.85 \bullet 10^{-3}$ kg of $H_2S$/kg of water).

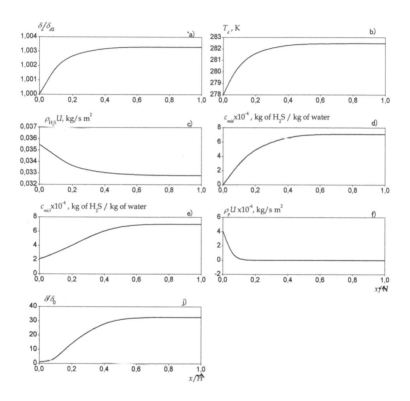

**Figure 7.** Calculation of combined air cleaning from submicron dust and hydrogen sulfide in direct-flow scrubber: $H$ =2 m; $q$ =0.015 m³/m³; $\delta_{d0}$ =700 μm; $V_{d0}$ =24.5 m/s; $U_0$ =0.25 m/s; $T_{d0}$ =278 K; $T_{00}$ =333 K; $d_0$ =0.2 kg/kg of dry air; $d_{H_2S,0}$ =0.2 kg/kg of dry air; $\rho_{p0}$ =1.72 g/m³; $\delta_0$ =1 μm; $\eta_{H_2S}$ =0.075411; $\eta_p$ =1.0

## 3.4. Calculation of absorption and condensation dust capture in Venturi scrubber

Calculation results on $H_2S$ absorption and condensation capture of dust with different sizes in Venturi scrubber are shown in Fig. 8. As an example the Venturi scrubber with following parameters was chosen for calculations: diameter of Venturi tube mouth $d_m$=0.02 m, diffuser length $l$=0.2 m, diffuser opening angle $\alpha$=6° ($\alpha$=6–7°, $l/d_m$=10–15 are recommended for nor-malized Venturi tube [6, 10]), vapor-gas flow velocity in the tube mouth $U_0$=80 m/s, initial velocity of droplets in the tube mouth $V_{d0}$=4 m/s, irrigation coefficient $q$=0.015 m³/m³, tem-perature of the vapor-gas flow and droplets in the tube mouth $T_{00}$=333 K and $T_{d0}$=278 K, re-spectively, concentration of dust particles at the inlet $\rho_{p0}$=1.72 g/m³, size of dust particles $\delta_0$=0.1 μm, moisture content in water vapor at the inlet was set $d_0$=0.2 kg/kg of dry air, gas content $d_{H_2S,0}$ =0.1 kg/kg of dry air. Efficient of $H_2S$ extraction and dust capture were deter-mined $\eta_{H_2S}$=0.072959 and $\eta_p$=0.52904, respectively.

The mean-mass size of droplets in the tube mouth was calculated by Nukiyama-Tanasava formula [1]:

$$\delta_{d0} = \frac{0,585}{U_0 - V_{d0}} \sqrt{\frac{\sigma_f}{\rho_f}} + 53,4 \left( \frac{\mu_f}{\sqrt{\rho_f \sigma_f}} \right)^{0,45} q^{1,5}, \text{ m,} \tag{66}$$

where $\rho_f$ (kg/m³), $\mu_f$ (Pa•s), $\sigma_f$ (N/m) and $q$ (m³/m³) are density, dynamic viscosity, surface tension coefficient of pneumatically atomized liquid, and irrigation coefficient.

Velocity $U$ was calculated with consideration of diffuser expansion angle [2, 11].

Dependences of droplet size along the diffuser length are presented in Fig. 8a). It can be seen that firstly condensation of water vapors occurs intensively, then this process stops at the length of $x/l$≈0.2, and the size of droplets stays constant up to the scrubber outlet. At this, the quantitative droplet size changes slightly along the diffuser length (it stays almost con-stant: the maximal increase is a little bit higher than 0.3 %).

A change in droplet temperature due to convective heat transfer between droplets and va-por-gas flow, thermal effects of water vapor condensation on droplets, and gas dissolution is shown in Fig. 8d). A change in mass concentration of $H_2S$ dissolved in a droplet is shown in Fig. 8c). It is obvious that absorption is almost completed at the length of tube diffuser 1 for this version of calculation. The same circumstance is illustrated by mass concentration of $H_2S$ in "formation" condensate along the diffuser in Fig. 8c). According to the figure, the sol-ubility limit on "formations" and droplets is not achieved as in the hollow jet scrubbers. A change in "formation" size due to water vapor condensate on their surfaces is illustrated in Fig. 8f). It can be seen that firstly water vapors condense very intensively, then at the dis-tance of about $x/l$≈0.1 this process completes, the size increases more than twice and stays constant until the leaving from the scrubber. Efficiency of dust capture in this version is up to 53 %. Calculation at the same parameters of the vapor-gas flow and dust at the scrubber inlet with mouth $d_z$=0.1 m and constrictor length $l$=1 m gives $\eta_p$=0.77729, $\eta_{H_2S}$=0.074965. It fol-lows from the diagrams in this figure that for the calculated version it is practically reasona-ble to be limited by diffuser length $x/l$≈0.4 ($x$=0.08 m), where the processes of dust capture (Fig. 8g)) and absorption are completed (Fig. 8b)). Therefore, the residual length of 0.12 m is excessive. Figs. 8b) and 8g) illustrate distributions of dust and $H_2S$ mass fluxes along the diffuser of Venturi tube.

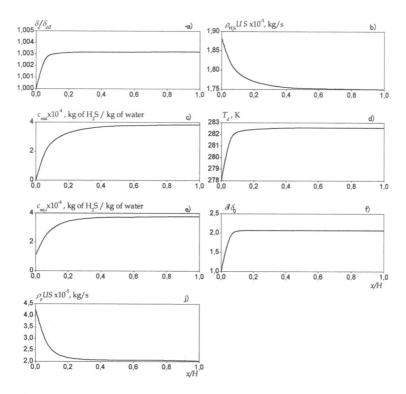

**Figure 8.** Calculation of $H_2S$ absorption and dust capture in Venturi scrubber (calculated parameters are presented in the text)

It is necessary to note that these calculation versions do not meet the conditions of optimal scrubber operation; they only illustrate the character of complex gas cleaning. To determine the optimal regimes, a series of calculation on the basis of suggested model should be carried out and analyzed for the specific industrial conditions.

Let's turn to comparison of calculation results with the known experimental data. The experimental volumetric mass transfer coefficient is shown in [6] for $NH_3$ absorption in the Venturi tube with mouth diameter $d_m$=0.02 m. There no geometrical and other parameters. This coefficient is presented as

$$\beta_{iv,e}^{NH_3} = 260 U_0^{1,56} q_l^{0,57},$$                                    (67)

where $q_i$ is irrigation coefficient in l/m³, $U_0$ is in m/s, and $\beta_{iv}$ is in 1/h.

The experimental value of H₂S absorbed in Venturi tube is expressed by formula, kg/h,

$$\Delta G_{i,e} = \beta_{iv,e}^{H_2S} \Delta \rho_{i,e} V_{dif},$$ (68)

where $V_{dif}$ is diffuser volume, $\Delta \rho_i$ is calculated drop of gas concentration along the diffuser length, kg/m³, corresponding to experimental data.

The theoretical value of mass of absorbed gas, kg/h, is

$$\Delta G_{i,th} = \beta_{iv,th}^{H_2S} \Delta \rho_{i,th} V_{dif} \approx \Delta \rho_{i,th} U_0 \frac{\pi d_m^2}{4} 3600.$$ (69)

Then

$$\frac{\Delta G_{i,th}}{\Delta G_{i,e}} = \frac{U_0 \frac{\pi d_m^2}{4} 3600 \Delta \rho_{i,th}}{\beta_{iv,e}^{NH_3} V_{dif} \left( \frac{D_{i,0}}{D_{NH_3,0}} \right)^{0,67} \Delta \rho_{i,e}},$$ (70)

where (see Fig. 9) the volume of truncated cone is

$$V_{dif} = \frac{\pi}{24} \frac{d_m^3}{tg \frac{\alpha}{2}} \left[ \left( \frac{2l}{d_m} tg \frac{\alpha}{2} + 1 \right)^3 - 1 \right].$$ (71)

Substituting (67), (71) into (70), at $\alpha$=6°, $l$=0.2 m, $d_m$=0.02 m, $U_0$=80 m/s, $q_i$=15 l/m³ we get the following, assuming that $\Delta \rho_{i,th} \approx \Delta \rho_{i,e}$,

$$\frac{\Delta G_{i,th}}{\Delta G_{i,e}} = 0,9457,$$ (72)

where for H₂S $\left( \frac{D_{i,0}}{D_{NH_3,0}} \right)^{0,67}$ =0.7426, i.e., the difference between the theory and experiment is less than 6 %, what is a good agreement, considering the assumed parameters for normalized Venturi tube.

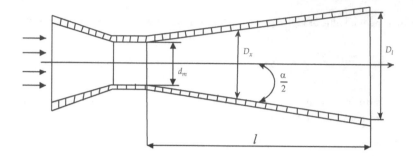

**Figure 9.** The scheme of Venturi tube

The amount of absorbed $H_2S$ for the version of calculation in Fig. 8 ($\eta_{H_2S}$=0.072959, $\rho_{d.a.,0}$ =0.7541 kg/m³) is

$$W_{i,th} = 0,7541 \cdot 0,1 \cdot 0,072959 \cdot 80 \frac{3,14(0,02)^2}{4} 3600 = 0,5 \text{ kg/h.}$$

For the scrubber with $d_m$=0.1 m, $l$=1 m at the same dust and gas parameters at the inlet $W_{i,e}$=12, 78 kg/h.

The experimental values of efficiency of condensation capture of submicron dust are compared with results of model calculation inn [2, 11, 12] at the example of deposition of ash particles from cracking gases under the industrial conditions in hollow jet scrubbers [13]; good agreement is achieved.

## 4. The choice of the value of calculated concentration difference for the absorbed gas component

Let's consider this important question in detail as an addition to iss. 2 at the example of water absorption of $SO_2$, comparing calculation and experimental data [6] on volumetric mass transfer coefficient.

It follows from equation (3) that

$$W_{i,th} = -\int_0^H \beta_{id} \pi \delta_{d0}^2 (\rho_{id} - \rho_i) n_d dx, \ kg/s \cdot m^2, \quad (73)$$

where, according to formulas (26) and (38)

$$n_d = 6q \frac{U_0}{V_{dx}} \frac{1}{\pi \delta_{d0}^3}. \tag{74}$$

In (73) and (74), according to calculation results, it is assumed that $\delta_d = \delta_{d0}$.

Let's put $\beta_{id}$ from (19) to (73) at $K_{ci} = 1$ and $n_d$ from (74), we obtain, proved by estimates,

0, $276 \text{Re}_d^{0,55} \text{Sc}_i^{0,33} \gg 1$, $V_{dx} \gg U$,

$$W_{i,th} \approx -3,312 \frac{qU_0}{\delta_{d0}} \int_0^H \frac{(\rho_{id} - \rho_i) \mathbf{dx}}{\text{Re}_d^{0,45} \text{Sc}_i^{0,67}}, \tag{75}$$

where

$$\text{Re}_d \approx \frac{V_{dx} \delta_{d0} \rho}{\mu}. \tag{76}$$

Lets' turn dependence (74) to the following form using the theorem about an average for integral:

$$W_{i,th} \approx 3,312 q U_0 \frac{\left(\bar{D}_{NH_3}\right)^{0,67}}{\delta_{d0}^{1,45} \bar{V}_d^{0,45}} \left(\bar{v}\right)^{-0,22} \left(\int_0^H \Delta \rho_{id} \mathbf{dx}\right) \left(\frac{D_{i,0}}{D_{NH_3,0}}\right)^{0,67}, \ kg/s \cdot m^2, \tag{77}$$

where $\bar{v} = (\bar{\mu} / \bar{\rho})$ is the average value of kinematic viscosity of the vapor-gas flow in the scrubber volume, m²/s, $\bar{D}_{NH_3}$ is the average value of diffusion coefficient of methane $NH_3$, $\bar{V}_{dx}$ is average velocity of droplets on the 0–H way at motion from scrubber inlet to the outlet, m/s, $\Delta \rho_{id} = | \rho_{id} - \rho_i |$.

Expressing velocity $U_0$ and $V_{dx}$ in m/h, and assuming $\delta_{d0} = 7 \cdot 10^{-4}$m (700 μm), $\bar{D}_{NH_3}(T_{d,ave} \approx 278, 5\,K) = 0, 198 \cdot 10^{-4} \left(\frac{285}{273}\right)^{1,75} = 0, 205 \cdot 10^{-4}$m²/s (see Fig. 10 b)), $\bar{v} = (285\,K) \approx 1, 5 \cdot 10^{-6}$ m²/s, $\bar{V}_{dx} \approx 5, 25$ m/s (see Fig. 10 e)), $\left(D_{SO_2,0} / D_{NH_3,0}\right)^{0,67} = 0, 715$, we obtain for calculation parameters of Fig. 10 $W_{i,th} \approx 83$ kg/h•m², where $\int_0^H \Delta \rho_{id} dx = 0, 6437$ kg/m² is obtained via model calculation (see Fig. 11).

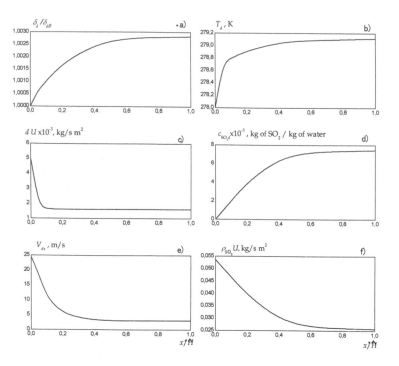

**Figure 10.** Calculation of SO$_2$ absorption in direct-flow jet scrubber: $H$ =12.75 m; $q$ =0.015 m³/m³; $\delta_{d0}$ =700 μm; $V_{d0}$ =24.5 m/s; $U_0$ =0.25 m/s; $T_{d0}$ =278 K; $T_{00}$ =293 K; d$_0$ =0.02 kg/kg of dry air; d$_{SO_2,0}$ =0.2 kg/kg of dry air; $\eta_{SO_2}$=0.51722

We should note that multiplier $(D_{i,0}/D_{NH_30})^{0,67}$ was included into formula (76) as a correction like to was dome for empirical dependence (46).

Numerical calculation by the model give the value of SO$_2$ extraction

$$W_{i,th} = \rho_{d.a.,0}d_{SO_2,0}U_0\eta_{SO_2} 3600 = 1,0743\cdot0,2\cdot\cdot0,25\cdot0,51722\cdot3600 = 100 \ \text{kg/h}\cdot\text{m}^2. \qquad (78)$$

The difference is 17 %, what is a sequence of simplifications and averaging in dependence (76).

If we assume average concentration difference in accordance to average experimental height $H$=7 m $\Delta\rho_i$=0, 6437/7=0, 092kg/m³, then

$$W_{i,e}=28685, 9U_0^{1,35}q^{0,45}H^{0,35}\Delta\rho_i\left(\frac{D_{i,0}}{D_{NH_30}}\right)^{0,67} =28685, 9\cdot0, 1539\cdot0, 151\cdot1, 976\cdot0, 092\cdot0, 715=86, 6kg/h\cdot m^2.$$

The difference with $W_{i,e}$=83 kg/h•m² is 4.2 %.

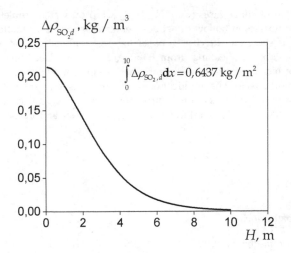

**Figure 11.** Distribution of $\Delta\rho_{SO_2,d}$ along the scrubber height for calculation parameters of Fig. 10.

If we take $\Delta\rho_i = \rho_{d.a.,0} d_{SO_2,0} \eta_{SO_2} = 0,111$ kg/m³, then $W_{i,e} = 104.5$ kg/h•m², what differs from result of (77) by similar 4.3 % with accuracy of estimation error. This proves the fact that calculated volumetric mass transfer coefficient agrees empirical dependence (46) of [6].

On the basis of analysis performed the calculated concentration difference should be recommended for practical application as the most appropriate

$$\Delta\rho_i = \rho_{i,0} - \rho_{i,H} \frac{U_H}{U_0} \qquad (79)$$

at determination of the value of extracted gas component by formula (49), thus, it is necessary to measure $\rho_{i,0}$, $\rho_{i,H}$ and $U_0$, $U_H$ at apparatus inlet and outlet. Calculation of moisture content is shown in Fig. 10 c), at this, it was obtained that $U_H = 0,2234$ m/s, $\rho_{d.a.,0} = 1,0743$ kg/m³.

## 5. Conclusions

The suggested physical-mathematical model of complex heat and mass transfer and condensation-absorption gas cleaning from dust and harmful gaseous components is confirmed by the known experimental data and can be used for engineering calculations and optimization of construction and operation parameters of hollow jet scrubbers of direct and counter flow types. This was proved by its numerical implementation for the specific conditions. Calculations on absorption of some gases ($CO_2$, $H_2S$, $SO_2$, $CH_4$) on water droplets, dispersed by coarse centrifugal nozzles in hollow direct-flow jet scrubber and pneumatic Venturi scrubber from wet air is shown in the current paper together with calculation of combined absorption-condensation air cleaning from $H_2S$ and various-sized fine dust in these apparatuses at an increased moisture content. The system of model equations is written at some certain conditions for the multicomponent vapor-gas mixture with particle. This makes it possible to use this system for calculation of complex gas cleaning from several harmful gas components and several fractions of dust particles and investigate regularities of this process.

## Nomenclature

$\vec{V}_d$ vector of droplet velocity

$\vec{U}$ nvector of vapor-gas low velocity

$\vec{g}$ vector of gravity acceleration

$m_d$ droplet mass (variable value due to evaporation-condensation and absorption), kg

$c_f$ specific heat capacity of liquid, J/ kg•K

$T_d$ mean mass temperature of droplets, K

$\alpha_d$ heat transfer coefficient of droplet, W/m²•K

$\delta_d$ mass-median size of droplet, m

$T$ temperature of vapor-gas flow, K

$r_i$ specific heat of absorption, evaporation-condensation, J/kg

$c_\delta$ specific heat capacity of "formation", J/kg•K

$\rho_\delta$ mass concentration of "formations' in the vapor-gas flow, kg/m³

$V_c = |\vec{V}_d - \vec{U}|$ module of relative droplet velocity, m/s

$\eta_{Stk}$ coefficient of "formation" capture by droplets

$T_\delta$ mean mass temperature of "formations", K

$\beta_{id}$ coefficient of droplet mass transfer with the i-th component of vapor-gas flow by concentration difference, m/s

$\rho_{id}$ and $\rho_i$ partial densities (mass concentrations) of saturated vapors of dissolvent and gas components near droplet surface and far from it (in the flow), kg/m³

$n_d$ and $n_\delta$ calculated concentrations of droplets and "formations" in the flow, $1/m^3$

$\rho_g$ partial density of non-reacting gas component, $kg/m^3$

$\rho_d$ mass concentration of droplets, $kg/m^3$

$\alpha_\delta$ heat transfer coefficient of "formation", $W/m^2 \bullet K$

$\delta$ size of "formation", m

$\rho$ density of vapor-gas flow, $kg/m^3$

$c$ specific heat capacity of vapor-gas flow, $J/kg \bullet K$

$\rho_p$ mass concentration of dry dust particles in the flow, $kg/m^3$

$\rho_f$ density of liquid (droplets), $kg/m^3$

$\mu$ dynamic viscosity of vapor-gas flow, $Pa \bullet s$

$M_i$ molar masses of components of the vapor-gas mixture, kg/kmole

$R = 8,314$ kJ/kmole$\bullet$K universal gas constant

$D_i$ diffusion coefficient of mixture component, $m^2/s$

$P_i$ partial pressure of the i-[th] component of the vapor-gas mixture, Pa

$P_{id,\delta}$ partial saturation pressures of mixture components, calculated by droplet and "formation" temperature, Pa

$m_{px,i}$ constants of phase equilibrium of solutions of i-[ths] components of extracted gases, Pa

$x_{id,\delta}$ mole fractions of gas components dissolved in a droplet and "formation" condensate

$c_{mid,\delta}$ mass fractions of gas components in droplet and "formation" dissolvent, kg/kg of dissolvent

$M_{dis}$ molar mass of dissolvent, kg/kmole

$dm_{v\delta}/d\tau$ rate of "formation" mass change due to evaporation-condensation of liquid, kg/s

$\delta_0$ initial size of dust particles, m

$\rho_s$ density of solution on "formation" due to condensation of liquid vapors and absorp-tion of gas components, $kg/m^3$

$d$ moisture content, kg of vapors/kg of dry non-reacting component of vapor-gas mixture

$d_i$ gas content, kg of reacting gas component/ kg of dry non-reacting component of vapor-gas mixture

$q = Q_f/Q_{sg0}$ irrigation coefficient

$Q_f$ volumetric flow rate of liquid, $m^3/s$

$Q_{sg0}$ volumetric flow rate of vapor-gas mixture at apparatus inlet, $m^3/s$

## Author details

M. I. Shilyaev and E. M. Khromova

*Address all correspondence to: helenka24@rambler.ru

Department of heating and ventilation, Tomsk State University of Architecture and Building, Tomsk, Russia

## References

[1] Shilyaev M.I., Shilyaev A.M., Grischenko E.P. Calculation Methods for Dust Catchers. Tomsk: Tomsk State University of Architecture and Building; 2006.

[2] Shilyaev M.I., Khromova E.M., Bogomolov A.R. Intensification of heat and mass transfer in dispersed media at condensation and evaporation. Tomsk: Tomsk State University of Architecture and Building; 2010.

[3] Shilyaev M.I., Khromova E.M. Simulation of heat and mass transfer in spray chambers. Theoretical Foundations of Chemical Engineering 2008; 42(4) 404-414.

[4] Tumasheva A.V. Modeling of heat and mass transfer processes in jet irrigation chambers: abstract of the thesis of candidate dissertation: 01.04.14: defended on 17.06.2011. Novosibirsk; 2011.

[5] Shilyaev M.I., Shilyaev A.M. Aerodynamics and heat and mass transfer of gas-dispersed flows. Tomsk: Tomsk State University of Architecture and Building; 2003.

[6] Ramm V.M. Absorption of Gases. Moscow: Khimiya; 1976.

[7] Pavlov K.F., Romankov P.G., Noskov A.A. Sums and Problems in the Course of Processes and Apparatuses of Chemical Technology. The 8th edition, revised and added. Leningrad: Khimiya; 1976.

[8] Perelman V.I. Brief Hand-Book of a Chemist. Edit. by Corr. Member of AS of USSR B.V. Nekrasov. The 3rd edition, revised and added. Moscow: State Scientific-Technical Publishing House of Chemical Literature; 1954.

[9] Goronovsly I.T., Nazarenko Yu.P., Nekryach E.F. Brief Chemistry Guide. Edit. by Academician of AS of USSR A.T. Pilipenko. Kiev: Naukova Dumka; 1987.

[10] Hand-Book on Dust and Ash Capture. Edit by A.A. Rusanov. Moscow: Energia; 1975.

[11] Shilyaev M.I., Khromova E.M. Capture of Fine Dust in Jet Scrubbers. In: Mohamed El-Amin (ed.) Mass Transfer in Multiphase Systems and its Applications. Vienna: In Tech4; 2011. 311-335.

[12] Shilyaev M.I., Khromova E.M., Grigoriev A.V., Tumashova A.V. Physical-mathematical model of condensation process of the sub-micron dust capture in sprayer scrubber. Thermophysics and Aeromechanics 2011; 18(3) 409-422.

[13] Uzhov V.N., Valdberg A.Yu. Gas Cleaning by Wet Filters. Moscow: Khimiya; 1972.

# Chapter 7

# Numerical Analysis of Mixed Convection Magnetohydrodynamic Heat and Mass Transfer past a Stretching Surface in a Micro-Polar Fluid-Saturated Porous Medium Under the Influence of Ohmic Heating

Sandile S. Motsa and Stanford Shateyi

Additional information is available at the end of the chapter

## 1. Introduction

Coupled head and mass transfer by mixed convection in a micro-polar fluid-saturated porous medium due to a stretching sheet has numerous applications in geophysics and energy related engineering problems that engineering problems that includes both metal and polymer sheets. The micro rotation of each particle about its centroid as well as the translatory motion of each particle are taken into account in the study of micro polar fluids.

Past studies on micro polar fluids include, among others, the boundary layer flow of a micro polar fluid over a plate (Rees and Bassom 1996), the flow of a micro polar fluid over a stretching sheet (Raptis,1998) and the flow of a micro polar of fluid in a porons medium (Rawat et al.2007; Motsa et al. 2010; Pal and Chatterjec 2011) .

In many engineering areas processes occur at high temperature so knowledge of radiation heat transfer plays very significant roles and cannot be neglected. Thermal radiation effects become important when the difference between the surface and the ambient temperature is large. Numerous studies have been made to analyze the effect of radiation boundary layer flows under different geometry, (Pal 2009, Pal and Mondal 2010; Shateyi and Motsa 2009; Shateyi and Motsa 2011; Pal and Chatterjec 2011), among others.

In view of the above discussions, we envisage to investigate the steady two-dimensional mixed convection and mass transfer flow past a semi infinite vertical porous plate embedded in a micro polar fluid-saturated porous medium in the presence of thermal radiation, Ohmic dissipation, inertia effects and dispersion effects as these parameters have significant contribution to convective transport process. The problem considered in this chapter has many practical situations such as polymer extrusion processes and the combined effects of the physical parameters will have a large impact on heat and mass transfer characteristics. In this chapter we also aim to solve the current problem using the successive linearization method (Motsa 2011; Motsa and Sibanda 2012) .

## 2. Mathematical formulation

We consider a steady two-dimensional magnetohydrodynamic laminar mixed convection heat and mass transfer flow of a viscous incompressible fluid over a vertical sheet in a micro polar fluid-saturated porous medium. A uniform transverse magnetic field $B_0$ is applied normal to the flow. In this chapter, we assume that the applied magnetic field is taken being weak so that Hall and ion-slip effects maybe neglected. The radiative heat flux in the x-direction is considered negligible in comparison to the y-direction. Under the usual boundary layer approximation, along with Boussinesq's approximations the governing equations describing the conservation of mass, momentum, energy and concentration in the presence of thermal radiation and ohmic heating are governed by the following equations:

$$\frac{\partial u}{\partial x} + \frac{\partial v}{\partial y} = 0, \tag{1}$$

$$u\frac{\partial u}{\partial x} + v\frac{\partial u}{\partial y} = \left(v + \frac{k_1^*}{\rho}\right)\frac{\partial^2 u}{\partial y^2} + \frac{k_1^*}{\rho}\frac{\partial N}{\partial y} - \left(\frac{v\varphi}{k} + \frac{\sigma B_0^2}{\rho}\right)u - \frac{C_b}{\sqrt{k}}\varphi u^2 + g\beta_t(T - T_\infty)$$
$$+ g\beta_c(C - C_\infty), \tag{2}$$

$$\rho j\left(u\frac{\partial N}{\partial x} + v\frac{\partial N}{\partial y}\right) = \gamma\frac{\partial^2 N}{\partial y^2} - k_1^*(2N + \frac{\partial u}{\partial y}), \tag{3}$$

$$u\frac{\partial T}{\partial x} + v\frac{\partial T}{\partial y} = \frac{1}{\rho c_p}\frac{\partial}{\partial y}\left(\kappa\frac{\partial T}{\partial y}\right) - \frac{1}{\rho c_p}\frac{\partial q_r}{\partial y} + \frac{\delta B_0^2}{\rho c_p}u^2 + \frac{\mu}{\rho c_p}\left(\frac{\partial u}{\partial y}\right)^2, \tag{4}$$

$$u\frac{\partial C}{\partial x} + v\frac{\partial C}{\partial y} = D\frac{\partial^2 C}{\partial y^2}. \tag{5}$$

Where $u, v$ are the velocity components along the x- and y - direction, $\rho$ is the density of the fluid, $T$ and $C$ are the temperature and concentration, respectively, $C_b$ is the form of drag coefficient which is independent of viscosity and other properties of the fluid but is dependent on the geometry of the medium, $k$ is the permeability of the porous medium, $\beta_t$ and $\beta_c$ are the coefficients of thermal and concentration expansions, respectively, $\gamma$ is the spin gradient and $k_1^*$ is the vortex viscosity, $c_p$ is the specific heat constant pressure, $v$ is the kinematic viscosity, $\sigma$ is the electrical conductivity of the fluid, $B_0$ is externally imposed magnetic field strength, $D$ is the molecular diffusivity , $j$ is the micro inertia per unit mass, $N$ is the component of microrotation or angular velocity whose rotation is in the $x - y$ plane direction. The spin gradient viscosity $\gamma$, defines the relationship between the coefficient of viscosity and micro-inertia as follows (Kim 1999):

$$\gamma = \mu\left(1 + \frac{K}{2}\right)j, \tag{6}$$

with $K = k_1^*/v$ being the material parameter. We take $j = v/b$ as a reference length. The thermal conductivity $\kappa$ is assumed to vary linearly with temperature and is of the form:

$$\kappa = \kappa_\infty[1 + \epsilon\theta(\eta)], \tag{7}$$

where $\epsilon$ is a small parameter, and following the Rosseland approximation, the radiative heat flux $q_r$ is modeled as,

$$q_r = -\frac{4\sigma^*}{3k^*}\frac{\partial T^4}{\partial y}, \tag{8}$$

where $\sigma^*$ is the Stefan-Boltzmann constant and $k^*$ is the mean absorption coefficient. We assume that the difference in temperature within the flow are such $T^4$ can be expressed as linear combination of the temperature and then $T^4$ can be expanded in Taylor's series about $T_\infty$ as follows:

$$T^4 = T_\infty^4 + 4T_\infty^3(T - T_\infty) + 6T_\infty^2(T - T_\infty)^2 + \dots\dots\dots \tag{9}$$

Now neglecting higher order terms beyond the first degree in $(T - T_\infty)$ gives

$$T^4 \cong -3T_\infty^4 + 4T_\infty^4 T. \tag{10}$$

Using equations (8) and (10) we obtain

$$\frac{\partial q_r}{\partial y} = -\frac{16T_\infty^3\sigma^*}{3k^*}\frac{\partial^2 T}{\partial y^2}, \tag{11}$$

Using equation (11) in equation (4) gives

$$u\frac{\partial T}{\partial x} + v\frac{\partial T}{\partial y} = \frac{1}{\rho c_p}\frac{\partial}{\partial y}\left(\kappa\frac{\partial T}{\partial y}\right) + \frac{16T_\infty^3\sigma^*}{3k^*}\frac{\partial^2 T}{\partial y^2} + \frac{\delta B_0^2}{\rho c_p}u^2 + \frac{\mu}{\rho c_p}\left(\frac{\partial u}{\partial y}\right)^2, \tag{12}$$

The appropriate boundary conditions for the problem under study are given by:

$$u = u_w = bx, \quad v = 0, \quad N = -n\frac{\partial u}{\partial y} \text{ at } y = 0, \quad u \to 0, \quad N \to 0 \text{ as } y \to \infty, \tag{13}$$

$$T = T_w = T_\infty + A_0\left(\frac{x}{l}\right)^2, \quad C = C_w = C_\infty + A_1\left(\frac{x}{l}\right)^2, \text{ at } y = 0, \tag{14}$$

where $A_0$, $A_1$ are constants, $l$ being the characteristics length, $T_w$ is the wall temperature of the fluid and $T_\infty$ is the ambient fluid temperature, $C_w$ is the wall concentration of the solute and $C_\infty$ is the concentration of the solute far away from the sheet, $n$ is a constant taken as $0 \le n \le 1$.

## 2.1. Similarity solutions

In order to reduce the governing equations into a convenient system of ordinary differential equations. We introduce the following self-similar solution of the form:

$$u = bxf'(\eta), \quad v = -\sqrt{bv}f(\eta), \quad \eta = \sqrt{\frac{b}{v}}y, \tag{15}$$

$$N = bx(b/v)^{\frac{1}{2}}g(\eta), \quad \theta(n) = \frac{T - T_\infty}{T_w - T_\infty}, \quad \phi(\eta) = \frac{C - C_\infty}{C_w - C_\infty}, \tag{16}$$

where $f$ is the dimensionless stream function and $\eta$ is the similarity variable. Substituting these into the governing equations, we obtain the following nonlinear ordinary differential equations:

$$(1 + K)f''' + ff'' - (1 + \alpha)f'^2 - \left(M^2 + \frac{1}{Da}\right)f' + Kg' + Gr_t\theta + Gr_m\phi = 0, \tag{17}$$

$$\left(1 + \frac{K}{2}\right)g'' + fg' - f'g - 2Kg - Kf'' = 0, \tag{18}$$

$$\frac{1}{Pr}(1 + R + \epsilon\theta)\theta'' + f\theta' + \frac{\epsilon}{Pr}(\theta')^2 - 2f'\theta + M^2E_c(f')^2 + Ec(f'')^2 = 0, \tag{19}$$

$$\frac{1}{Sc}\phi'' + f\phi' - 2\phi f' = 0, \tag{20}$$

where $\alpha = \frac{C_b}{\sqrt{k}}\varphi x$ is the local inertia coefficient parameter, $\frac{1}{Da} = \frac{\varphi r}{Kb}$ is inverse Darcy number, $M = \sqrt{\frac{\delta}{\rho b}}B_0$ is the Hartmann number, $Gr_t = \frac{g\beta_t(T-T_\infty)}{b^2l}$ is the local Grashof number, $Gr_c = \frac{g\beta_c(C-C_\infty)}{b^2l}$ local concentration Grashof number and $K = k^*/v$ is the material parameter, $Pr = \frac{\mu c_p}{k_\infty}$ is the Prandtl number, $E_c = \frac{b^2l^2}{Ac_p}$ is Ekert number, $R = \frac{16\delta^*T_\infty^3}{3k_\infty k^*}$ is the thermal radiation parameter, $S_c = \frac{v}{D}$ is the Schmidt number. The appropriate boundary conditions 12 and 13 now become:

$$f(\eta) = 0, \ f'(\eta) = 1, \ g(\eta) = -nf''(\eta), \ \theta(\eta) = 1, \ \phi(\eta) = 1, \ \text{at } \eta = 0, \tag{21}$$

$$f'(\eta) \to 0, \ g(\eta) \to 0, \ \theta(\eta) \to 0 \ \phi(\eta) \to 0 \ \text{as} \to \infty. \tag{22}$$

## 3. Method of solution

The governing nonlinear problem (16 - 19) is solved using the successive linearization method (SLM). In its basic form, the SLM ([4, 5]) seeks to linearize the governing nonlinear differential equations to a system of linear differential equations which, in most cases, cannot be solved analytically. The Chebyshev pseudospectral method (or any other collocation method or numerical scheme) is then used to transform the iterative sequence of linearized differential equations into a system of linear algebraic equations.

To solve the system of nonlinear equations (16 - 19), we introduce the following notation,

$$(z_1, z_2, z_3, z_4) = (f, g, \theta, \phi) \tag{23}$$

to represent the governing independent variables. In terms of the variables (23), we define the following vector of derivatives of $z_j$ ($j = 1, 2, 3, 4$), with respect to $\eta$

$$Z_1 = [f, f', f'', f'''] = [z_1^{(0)}, z_1^{(1)}, z_1^{(2)}, z_1^{(3)}], \tag{24}$$

$$Z_2 = [g, g', g''] = [z_2^{(0)}, z_2^{(1)}, z_2^{(2)}], \tag{25}$$

$$Z_3 = [\theta, \theta', \theta''] = [z_3^{(0)}, z_3^{(1)}, z_3^{(2)}], \tag{26}$$

$$Z_4 = [\phi, \phi', \phi''] = [z_4^{(0)}, z_4^{(1)}, z_4^{(2)}]. \tag{27}$$

In general we have

$$Z_i = [z_i^{(0)}, z_i^{(1)}, \ldots, z_i^{(n_i)}], \tag{28}$$

where $z_i^{(0)} = z_i$, $z_i^{(p)}$ is the $p$th derivative of $z_i$ with respect to $\eta$ and $n_i$ ($i = 1, 2, \ldots m$) is the highest derivative order of the variable $z_i$ appearing in the system of equations. The system (16 - 19) can be written as a sum of it's linear $\mathcal{L}$ and nonlinear components $N$ as

$$\mathcal{L}[z_1(\eta), z_2(\eta), z_3(\eta), z_4(\eta)] + \mathcal{N}[z_1(\eta), z_2(\eta), z_3(\eta), z_4(\eta)] = 0 \tag{29}$$

subject to the boundary conditions

$$A_i[z_1(0), z_2(0), z_3(0), z_m(0)] = K_{a,i}, \quad B_b[z_1(\infty), z_2(\infty), z_3(\infty), z_4(\infty)] = K_{b,i}, \tag{30}$$

where $A_i$ and $B_i$ are linear operators and $K_{a,i}$ and $K_{b,i}$ are constants for $i = 1, 2, \ldots, 4$. In addition, we define $L_i$ and $N_i$ to be the linear and nonlinear operators, respectively, that operate on the $Z_i$ for $i = 1, 2, 3, 4$ With these definitions, equation (29) and (30) can be written as

$$L_i[Z_1, Z_2, Z_3, Z_4] + N_i[Z_1, Z_2, Z_3, Z_4] = \sum_{j=1}^{4} \sum_{p=0}^{n_j} \alpha_{i,j}^{[p]} z_j^{(p)} + N_i[Z_1, Z_2, Z_3, Z_4] = 0 \tag{31}$$

where $\alpha_{i,j}^{[p]}$ are the constant coefficient of $z_j^{(p)}$, the derivative of $z_j$ ($j = 1, 2, 3, 4$) that appears in the $i$th equation for $i = 1, 2, 3, 4$

The boundary conditions (30) can be written as

$$\sum_{j=1}^{4} \sum_{p=0}^{n_j-1} \beta_{v,j}^{[p]} z_j^{(p)}(0) = K_{a,v}, \quad v = 1, 2, \ldots, m_a \tag{32}$$

$$\sum_{j=1}^{4} \sum_{p=0}^{n_j-1} \gamma_{\sigma,j}^{[p]} z_j^{(p)}(\infty) = K_{b,\sigma}, \quad \sigma = 1, 2, \ldots, m_b \tag{33}$$

where $\beta_{v,j}^{[p]}$ ($\gamma_{\sigma,j}^{[p]}$) are the constant coefficients of $z_j^{(p)}$ in the boundary conditions, and $m_a$, $m_b$ are the total number of prescribed boundary conditions at $\eta = 0$ and $\eta = \infty$ respectively. We remark that the sum $m_a + m_b$ is equal to the sum of the highest orders of the derivatives corresponding to the dependent variables $z_i$, that is

$$m_a + m_b = \sum_{i=1}^{m} n_i. \tag{34}$$

The SLM assumes that the solution of (29) can be obtained as the convergent series

$$z_i(x) = \sum_{r=0}^{+\infty} z_{i,r}, \tag{35}$$

which, for numerical implementation is truncated at $r = s$ and written as

$$z_i(x) = z_{i,s} + \sum_{r=0}^{s-1} z_{i,r}. \tag{36}$$

A recursive iteration scheme is obtained by substituting (36) into the governing equation (29) and linearizing by neglecting nonlinear terms in $z_{i,s}$ and all it's derivatives. Substituting (36) in (29 - 30) and linearizing gives

$$L_i[Z_{1,s}, Z_{2,s}, Z_{3,s}, Z_{4,s}] + \sum_{j=0}^{4} \sum_{p=0}^{n_j} z_{j,s}^{(p)} \frac{\partial N_i}{\partial z_j^{(p)}} [....] = -L_i[....] - N_i[....], \tag{37}$$

subject to

$$\sum_{j=1}^{4} \sum_{p=0}^{n_j-1} \beta_{v,j}^{[p]} z_{j,s}^{(p)}(0) = 0, \quad v = 1, 2, \ldots, m_a \tag{38}$$

$$\sum_{j=1}^{4} \sum_{p=0}^{n_j-1} \gamma_{\sigma,j}^{[p]} z_{j,s}^{(p)}(\infty) = 0, \quad \sigma = 1, 2, \ldots, m_b \tag{39}$$

where

$$[....] = \left[\sum_{r=0}^{s-1} Z_{1,r}, \sum_{r=0}^{s-1} Z_{2,r}, \sum_{r=0}^{s-1} Z_{3,r}, \sum_{r=0}^{s-1} Z_{4,r}\right] \tag{40}$$

The initial approximation $,z_{j,0}(\eta)$, required to start the iteration scheme (37) is chosen to be a function that satisfies the boundary conditions (30). As a guide, the initial guess can be obtained as a solution of the linear part of (29) subject to the boundary conditions (30), that is, we solve

$$\sum_{j=1}^{m} \sum_{p=0}^{n_j} \alpha_{i,j}^{[p]} z_{j,0}^{(p)} = 0 \tag{41}$$

subject to

$$\sum_{j=1}^{4} \sum_{p=0}^{n_j-1} \beta_{v,j}^{[p]} z_{j,0}^{(p)}(0) = K_{a,v}, \qquad v = 1,2,\ldots,m_a \tag{42}$$

$$\sum_{j=1}^{4} \sum_{p=0}^{n_j-1} \gamma_{\sigma,j}^{[p]} z_{j,0}^{(p)}(\infty) = K_{b,\sigma}, \qquad \sigma = 1,2,\ldots,m_b \tag{43}$$

To solve the iteration scheme (37), it is convenient to use the Chebyshev spectral collocation method. For brevity, we omit the details of the spectral methods, and refer interested readers to [[12, 13]]. Before applying the spectral method, it is convenient to transform the domain on which the governing equation is defined to the interval [-1,1] on which the spectral method can be implemented. We use the transformation $\eta = \eta_e(\tau+1)/2$ to map the interval $[0,\eta_e]$ to [-1,1]. Here, it is assumed that $\eta_e$ is a finite real number which is chosen to be large enough to numerically approximate infinity. The basic idea behind the spectral collocation method is the introduction of a differentiation matrix $D$ which is used to approximate the derivatives of the unknown variables $z_i(\eta)$ at the collocation points as the matrix vector product

$$\frac{dz_i}{d\eta} = \sum_{k=0}^{\bar{N}} \mathbf{D}_{lk} z_i(\tau_k) = \mathbf{D}\mathbf{Z}_i, \qquad l = 0,1,\ldots,\bar{N} \tag{44}$$

where $\bar{N}+1$ is the number of collocation points (grid points), $\mathbf{D} = 2D/\eta_e$, and $\mathbf{Z} = [z(\tau_0), z(\tau_1), \ldots, z(\tau_N)]^T$ is the vector function at the collocation points. Higher order derivatives are obtained as powers of $\mathbf{D}$, that is

$$z_j^{(p)} = \mathbf{D}^p \mathbf{Z}_j. \tag{45}$$

Applying the Chebyshev spectral collocation on the recursive iteration scheme (37) gives

$$\sum_{j=1}^{m} \left[ \Lambda_{i,j} + \Pi_{i,j} \right] \mathbf{Z}_{j,s} = \Phi_{i,s-1}, \quad i,j = 1,2,3,4 \tag{46}$$

where $\mathbf{Z}_{i,s} = [z_{i,s}(\tau_0), z_{i,s}(\tau_1), \ldots, z_{i,s}(\tau_N)]^T$, $\Lambda_{i,j}$, $\Pi_{i,j}$ and $\Phi_i$ are given by

$$\Lambda_{i,j} = \sum_{p=0}^{n_j} \alpha_{i,j}^p \mathbf{D}^p, \quad \Pi_{i,j} = \sum_{p=0}^{n_j} \frac{\partial N_i}{\partial z_j^{(p)}} \mathbf{D}^p, \quad i,j = 1,2,3,4. \tag{47}$$

and

$$\Phi_{i,s-1} = -L_i[\ldots] - N_i[\ldots], \tag{48}$$

respectively.

Defining $\Delta = \Lambda + \Pi$, we can write equation (46) in matrix form as

$$\begin{bmatrix} \Delta_{1,1} & \Delta_{1,2} & \cdots & \Delta_{1,m} \\ \Delta_{2,1} & \Delta_{2,2} & \cdots & \Delta_{2,m} \\ \vdots & \vdots & & \vdots \\ \Delta_{m,1} & \Lambda_{m,2} & \cdots & \Lambda_{m,m} \end{bmatrix} \begin{bmatrix} \mathbf{Z}_{1,s} \\ \mathbf{Z}_{2,s} \\ \mathbf{Z}_{3,s} \\ \mathbf{Z}_{4,s} \end{bmatrix} = \begin{bmatrix} \Phi_{1,s-1} \\ \Phi_{2,s-1} \\ \Phi_{3,s-1} \\ \Phi_{4,s-1} \end{bmatrix} \tag{49}$$

where $\mathbf{Z}_{i,s}$, $\Phi_{i,s-1}$ are vectors of size $(\bar{N}+1) \times 1$ and $\Delta_{i,j}$ are $(\bar{N}+1) \times (\bar{N}+1)$ matrices . After imposing the boundary conditions on the matrix system (49), and starting from $Z_{i,0}$, the recursive sequence (49) is solved iteratively for $s = 1,2,3\ldots$ and the approximate solution for each $z_i(\eta)$ is obtained from the series

$$z_i(\eta) = z_{i,0}(\eta) + z_{i,1}(\eta) + z_{i,2}(\eta) + z_{i,3}(\eta) + \cdots \tag{50}$$

## 4. Results and discussion

This section presents the effects of various parameters on the velocity, temperature, microrotation and concentration profiles. We remark that the, unless otherwise specificd, the SLM results presented in this analysis where obtained using $N = 100$ collocation points and $\eta_e = 20$ was used as a numerical approximation infinity. In order to get physical insight into the problem, the effects of these parameters encountered in the governing equations of the problem are analyzed with the help of figures. Figures 1 to 3 depicts effects of the local inertia coefficient parameter $\alpha$, on the velocity, concentration and microrotation distributions. From Figure 1 it is observed that the horizontal velocity profiles decrease with the increasing values of $\alpha$ due to the fact that the second-order quadratic drag is offered by the porous medium to the fluid motion. This drag force results in decreased fluid in the boundary layer. In turn the reduced fluid flow causes the concentration $\phi(\eta)$ to increase as depicted in Figure 2. As expected the increasing values of the local inertia coefficient parameter $\alpha$, causes the gyration component $g(\eta)$ as shown in Figure 3.

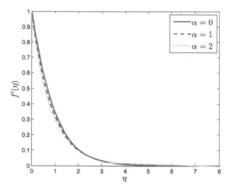

**Figure 1.** Variation of $f'(\eta)$ for different values of $\alpha$ with $K = 0.2; Da = 1; Grt = 1; M = 1; Grm = 1; \ Pr = 0.71; R = 1; \epsilon = 0.01; Ec = 1; Sc = 1; n = 0.5$.

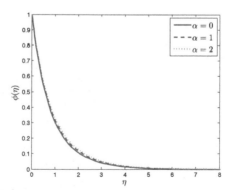

**Figure 2.** Plot of $\phi(\eta)$ for different values of $\alpha$ with $K = 0.2; Da = 1; Grt = 1; M = 1; Grm = 1; \ Pr = 0.71; R = 1; \epsilon = 0.01; Ec = 1; Sc = 1; n = 0.5$.

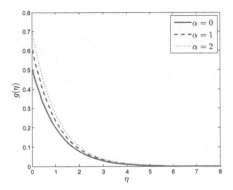

**Figure 3.** Variation of $g(\eta)$ for different values of $\alpha$ with $K = 0.2; Da = 1; Grt = 1; M = 1; Grm = 1; Pr = 0.71; R = 1; \epsilon = 0.01; Ec = 1; Sc = 1; n = 0.5$.

The effects of thermal Grashof number $Gr_t$ on the velocity, temperature, concentration and microrotation distributions are displayed in Figure 4 through Figure 7. From Figure 4 we observe that the horizontal velocity profiles increase with increasing values of the thermal Grashof number $Gr_t$. Buoyancy force acts like a favourable pressure gradient which in turn accelerates the fluid flow within the boundary layer. This accelerated fluid flow leads to the reduction of both the fluid temperature and concentration as can be seen from Figure 5 and Figure 6, respectively. The microrotation profiles are significantly affected by the thermal buoyancy parameter as shown in Figure 7. The increasing values of $Gr_t$ causes the microrotation significantly decrease.

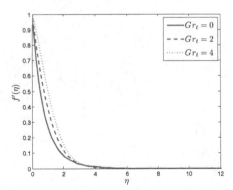

**Figure 4.** The influence of $Gr_t$ on $f'(\eta)$ with $K = 0.2; \alpha = 0.5; Da = 1; Mm = 1; Grm = 1; Pr = 0.71; R = 1; \epsilon = 0.01; Ec = 1; Sc = 1; n = 0.5$.

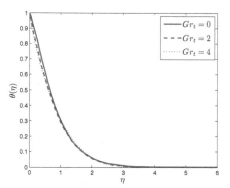

**Figure 5.** Variation of $Gr_t$ on $\theta(\eta)$ when $K = 0.2; \alpha = 0.5; Da = 1; Mm = 1; Grm = 1; Pr = 0.71; R = 1; \epsilon = 0.01; Ec = 1; Sc = 1; n = 0.5$.

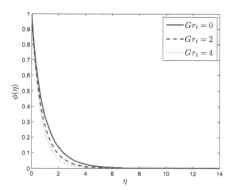

**Figure 6.** $Plot of \phi(\eta)$ for various values of $Gr_t$ when $K = 0.2; \alpha = 0.5; Da = 1; Mm = 1; Grm = 1; Pr = 0.71; R = 1; \epsilon = 0.01; Ec = 1; Sc = 1; n = 0.5$.

Numerical Analysis of Mixed Convection Magnetohydrodynamic Heat and Mass Transfer past a
Stretching Surface in a Micro-Polar Fluid-Saturated Porous Medium Under the Influence of Ohmic Heating

185

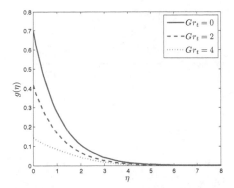

**Figure 7.** Plot of $g(\eta)$ when varying $Gr_t$ with $K = 0.2; \alpha = 0.5; Da = 1; Mm = 1; Grm = 1; Pr = 0.71; R = 1; \epsilon = 0.01; Ec = 1; Sc = 1; n = 0.5$.

Figures 8-11 display results of velocity, temperature, concentration and microrotation distribution for various values of the magnetic parameter $M$. As expected, the existence of the magnetic field is to decrease the velocity in the boundary layer because the application of the transverse magnetic field results in a drag type of force known as Lorentz force. The drag force resists the fluid flow which results in reducing the velocity of the fluid in the boundary layer. The temperature in the boundary layer increases with increasing values of $M$, as shown in Figure 9. From Figure 10, we also observe that the concentration distributions increase as the magnetic parameter $M$ increases. As the magnetic field is applied in the direction of the gyration component, the increasing values of $M$ cause the microrotation distribution to increase as depicted in Figure 11.

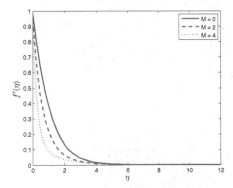

**Figure 8.** The influence of $M$ on $f'(\eta)$ with $K = 0.2; \alpha = 0.5; Da = 1; Grt = 1; Grm = 1; Pr = 5; R = 1; \epsilon = 0.01; Ec = 1; Sc = 1; n = 0.5$.

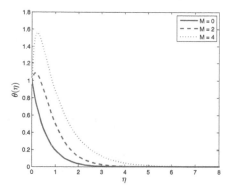

**Figure 9.** Plot of $\theta(\eta)$ for various values of $M$ with $K = 0.2; \alpha = 0.5; Da = 1; Grt = 1; Grm = 1; Pr = 5; R = 1; \epsilon = 0.01; Ec = 1; Sc = 1; n = 0.5$.

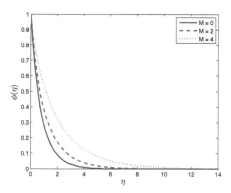

**Figure 10.** Variation of $\phi(\eta)$ for different values of $M$ with $K = 0.2; \alpha = 0.5; Da = 1; Grt = 1; Grm = 1; Pr = 5; R = 1; \epsilon = 0.01; Ec = 1; Sc = 1; n = 0.5$.

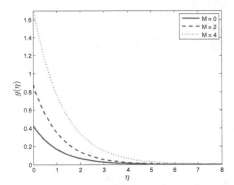

**Figure 11.** Plot of $g(\eta)$ when varying $M$ when $K = 0.2; \alpha = 0.5; Da = 1; Grt = 1; Grm = 1; Pr = 5; R = 1; \epsilon = 0.01; Ec = 1; Sc = 1; n = 0.5$.

The influence of the Darcy number $Da$ on the velocity, temperature, concentration and microrotation profiles is depicted in Figure 12-15. It is clearly observed from these figures that the velocity distribution $f'(\eta)$ increasing with increasing values of the Darcy number $Da$, whereas revers trend is seen on the temperature, concentration and microrotation distributions. This is because the presence of porous medium is to increase the resistance to the flow which causes the fluid flow to decrease.

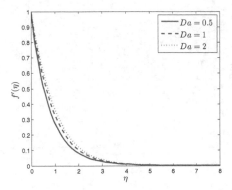

**Figure 12.** Variation of $f'(\eta)$ for different values of $Da$ with $K = 0.2; \alpha = 0.5; Grt = 1; Mm = 1; Grm = 1; Pr = 5; R = 1; e = 0.01; Ec = 1; Sc = 1; n = 0.5$.

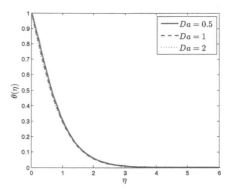

**Figure 13.** Plot of $\theta(\eta)$ for different values of $Da$ $K = 0.2; \alpha = 0.5; Grt = 1; Mm = 1; Grm = 1; Pr = 5; R = 1; e = 0.01; Ec = 1; Sc = 1; n = 0.5$.

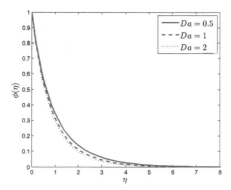

**Figure 14.** Influence of $Da$ on $\phi(\eta)$ when $K = 0.2; \alpha = 0.5; Grt = 1; Mm = 1; Grm = 1; Pr = 5; R = 1; e = 0.01; Ec = 1; Sc = 1; n = 0.5$.

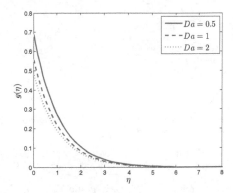

**Figure 15.** The effect of $Da$ on $g(\eta)$ when $K = 0.2; \alpha = 0.5; Grt = 1; Mm = 1; Grm = 1; Pr = 5; R = 1; e = 0.01; Ec = 1; Sc = 1; n = 0.5.$

## 5. Conclusion

Numerical analysis has been carried out in this chapter to study mixed convection heat and mass transfer in MHD flow past a stretching sheet in a micropolar fluid saturated medium under the influence of Ohmic heating. The governing partial differential equations which describe the problem are transformed in a system of ordinary differential equations by using suitable similarity transformations. A recently developed iterative technique together with Chebyshev spectral collocation method is used to solve the highly non-linear and coupled ordinary differential equations. The effects of various physical parameters on the velocity, microrotation, temperature and concentration are obtained. The following main conclusions can be drawn from the present study:

1. The fluid velocity increases with increase in the Grashof numbers, but decreases with increasing values of the Hartman number, local inertia coefficient parameter and inverse Darcy number.

2. The microrotation profiles decrease with thermal/solutal buoyancy force, and the Darcy number, whereas opposite trends are seen by enhancing values of the magnetic field and inertia coefficient parameter.

3. The fluid temperature increases with increasing values of magnetic field, inertia coefficient parameter, while opposite effects are seen by enhancing buoyancy forces and Darcy number.

4. Concentration decreases with increasing values of buoyancy forces and the Darcy number, whereas reverse trends are seen with increasing values of magnetic field, inertia coefficient parameter.

## Acknowledgement

The authors wish to acknowledge financial support from the National Research Foundation (NRF).

## Author details

Sandile S. Motsa[1] and Stanford Shateyi[2]

1 University of KwaZulu-Natal, South Africa
2 University of Venda, South Africa

## References

[1] D. A. S. Rees, and A. P. Bassom, The Blasius Boundary Layer Flow of a Micropolar Fluid, Int. J. Eng. Sci. 34, 113-124 (1996).

[2] A. Raptis, Flow of a Micropolar Fluid Past a Continuously Moving Plate by the Presence of Radiation, Int. J. Heat Mass Transf. 41, 2865–2866 (1998).

[3] S. Rawat, R. Bhargava and O. A. Bŕeg, A Finite Element Study of the Transport Phenomena in MHD Micropolar Flow in a Darcy-Forchheimer Porous Medium, Proc. WCECS, San Francisco, CA (2007).

[4] S.S. Motsa, New algorithm for solving non-linear BVPs in heat transfer, International Journal of Modeling, Simulation & Scientific Computing, 2:3, 355–373, (2011)

[5] Motsa,S.S.,Sibanda,P. (2012). A linearisation method for non-linear singular boundary value problems, Computers and Mathematics with Applications, 63,1197–1203 (2012)

[6] S.S Motsa,S. Shateyi and P. Sibanda, A Model of Steady Viscous Flow of a Micropolar Fluid Driven by Injection or Suction Between a Porous Disk and a Non-Porous Disk Using a Novel Numerical Technique, Can. J. Chem. Eng. 88:991–1002, (2010).

[7] D. Pal and S. Chatterjee, Mixed convection magnetohydrodynamic heat and mass transfer past a stretching surface in a micropolar fluid-saturated porous medium under the influence of Ohmic heating, Soret and Dufour, Comm. Nonlinear Sci. Numer. Simul. 16, 1329-1346, (2011).

[8] D. Pal, Heat and mass transfer in stagnation point flow towards in a stretching surface in the presence of buoyancy force and thermal radiation, Meccanica, 44,145-158, (2009).

[9] Pal, D., and Mondal, H, The influence of thermal radiation on hydromagnetic Darcy-Forchheimer mixed convection flow past a stretching sheet embedded in a porous medium, Meccanica, 45(2), doi 10.1007/s11012- 010-9334-8, (2010).

[10] S. Shateyi , Motsa S.S,Thermal Radiation Effects on Heat and Mass Transfer over an Unsteady Stretching Surface, Mathematical Problems in Engineering Volume 2009, Article ID 965603, 13 pages doi:10.1155/2009/965603.

[11] S. Shateyi, and Motsa, S. S. Hydromagnetic non-Darcy flow, heat and mass transfer over a stretching sheet in the presence of thermal radiation and Ohmic dissipation. The Canadian Journal of Chemical Engineering, 89: n/a. doi: 10.1002/cjce.20499,(2011).

[12] C. Canuto, M.Y. Hussaini, A. Quarteroni, and T.A. Zang, Spectral Methods in fluid dynamics, Springer-Verlag, Berlin, 1988.

[13] L. N. Trefethen, Spectral Methods in MATLAB, SIAM, 2000.

# Chapter 8

# Fouling in Membrane Filtration and Remediation Methods

A. Abdelrasoul, H. Doan and A. Lohi

Additional information is available at the end of the chapter

## 1. Introduction

The growth of the membrane technologies has fell far behind the initial anticipation, one of the major obstacles, which hinders more widespread of its application, is that the filtration performance inevitably decreases with filtration time. This phenomenon is commonly termed as membrane fouling, which refers to the blockage of membrane pores during filtration by the combination of sieving and adsorption of particulates and compounds onto the membrane surface or within the membrane pores. Pore blockage reduces the permeate production rate and increases the complexity of the membrane filtration operation. This is the most challenging issue for further membrane development and applications.

Permeate flux and transmembrane pressure (TMP) are the best indicators of membrane fouling. Membrane fouling leads to a significant increase in hydraulic resistance, manifested as permeate flux decline or TMP increase when the process is operated under constant-TMP or constant-flux conditions. In a system where the permeate flux is maintained by increasing TMP, the energy required to achieve filtration increases. Over a long period of operation, membrane fouling is not totally reversible by backwashing. As the number of filtration cycles increases, the irreversible fraction of membrane fouling also increases. In order to obtain the desired production rate, chemical cleaning is required for membrane to regain most of its permeability. The resultant elevated cost makes membranes economically less feasible for many separation processes. There are also concerns that repeated chemical cleaning might affect the membrane life.

Fouling can be broadly classified into backwashable or non-backwashable, and reversible or irreversible based on the attachment strength of particles to the membrane surface. Backwashable fouling can be removed by reversing the direction of permeate flow through the pores of the membrane at the end of each filtration cycle. Non-backwashable fouling is the fouling that cannot be removed by normal hydraulic backwashing in between filtration cycles. However, non-backwashable fouling of the membrane can be handled by chemical cleaning. On the other hand, irreversible fouling cannot be removed with flushing, backwashing, chemical cleaning, or any other means, and the membrane cannot be restored to its original flux. Fouling also can be classified, based on the type of fouling material, into four categories: inorganic fouling/scaling, particle/colloidal fouling, microbial/biological fouling, and organic fouling. Inorganic fouling or scaling is caused by the accumulation of particles when the concentration of the chemical species exceeds its saturation concentration. Several studies have shown that increased concentration of $Ca^{2+}$ and $Mg^{2+}$ caused more fouling [1-3]. On the other hand, organic fouling occurs due to the clogging of the membrane by organic substances, and organic carbons generally concentrate on the internal surface of the membrane [4]. Based on the analysis of the extracted solution during chemical cleaning, it was found that most soluble organic foulants were of low molecular weights, and calcium was the major inorganic foulant [5].

Natural organic matter (NOM) is the organic material present in surface or ground water and contains various high molecular weight organic compounds. NOM includes both humic and non-humic fractions. The humic fraction consists of high molecular weight organic molecules. Common non-humic NOM foulants are proteins, amino sugars, polysaccharides, and polyoxyaromatics [6]. Several studies have shown that NOM is the major ultrafiltration membrane foulant, and different components of NOM cause different forms of fouling [7-9]. According to Makdissy *et al.*, the organic colloidal fraction causes significant fouling [10]. However, polysaccharides are identified as the dominant foulant [11]. Other studies reported that most fouling was caused by hydrophobic NOM components [12]. Nevertheless, neutral hydrophilic NOM components were found the major foulants by some researchers [13]. The NOM components, as the major foulants, can be ranked in the order neutral hydrophilics > hydrophobic acids > transphilic acids > charged hydrophilics. Due to conflicting results from different researchers and many facets of membrane fouling, there would be no universal solution for membrane fouling remediation, but it has to be dealt with and designed specifically for a certain type of foulant and membrane in use, as presented later in this paper.

## 2. Membrane fouling mechanism

A typical flux-time curve of ultrafiltration (UF), as shown in Figure 1, starts with (I) a rapid initial drop of the permeate flux, (II) followed by a long period of gradual flux decrease, and (III) ended with a steady-state flux.

Flux decline in membrane filtration is a result of the increase in the membrane resistance by the membrane pore blockage and the formation of a cake layer on the membrane surface. The pore blocking increases the membrane resistance while the cake formation creates an additional layer of resistance to the permeate flow. Pore blocking and cake formation can be considered as two essential mechanisms for membrane fouling.

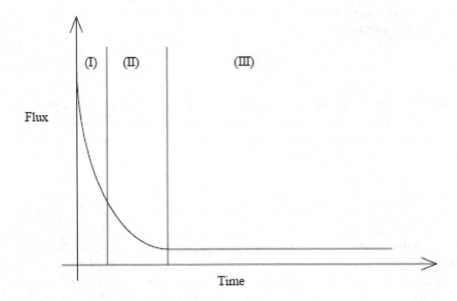

**Figure 1.** A Schematic presentation of the three stages in flux decline [14]

The rapid initial drop of the permeate flux can be attributed to quick blocking of membrane pores. The maximal permeate flux always occurs at the beginning of filtration because membrane pores are clean and opened at that moment. Flux declines as membrane pores are being blocked by retained particles. Pores are more likely to be blocked partially and the degree of pore blockage depends on the shape and relative size of particles and pores. The blockage is generally more complete when the particles and pores are similar in both shape and size [15-17]. Pore blocking is a quick process compared with cake formation since less than one layer of particles is sufficient to achieve the full blocking [16, 18].

Further flux decline after pore blockage is due to the formation and growth of a cake layer on the membrane surface. The cake layer is formed on the membrane surface as the amount of retained particles increases. The cake layer creates an additional resistance to the permeate flow and the resistance of the cake layer increases with the growth of cake layer thickness. Consequently, the permeate flux continues decreasing with time.

## 3. Mathematical models for membrane fouling

Pursuant to the understanding of different roles of aquatic components in membrane fouling, different mathematical models have been developed to describe the membrane fouling. The most widely used empirical model is the cake filtration model that focuses on the role of particles larger than membrane pore sizes. In this model, the hydrodynamic resistance of cake layer [Rc, m$^{-1}$] is defined as:

$$R_c = \hat{R}_c.m_d \tag{1}$$

where $\hat{R}_c$ [m/kg] is the specific cake resistance of the cake layer on the membrane surface and $md$ [kg/m$^2$] is the mass of deposit per unit surface area of membrane. The corresponding permeate flux (J, m$^3$/m$^2$.s) is expressed using Darcy's law and a resistance-in-series model (RIS) as below:

$$J = \frac{\Delta P}{\mu\ (R_m + R_c)} \tag{2}$$

where $\Delta P$ (Pa) is transmembrane pressure, $\mu$ (Pa-s) is the solution viscosity and $Rm$ (1/m) is the hydrodynamic resistance of clean membrane. Additional work has been done to relate $\hat{R}_c$ to the structure of the cake layer formed by particles or aggregates [19, 20]. The cake filtration model has been used to fit filtration data and reasonable results have been obtained [21]. However, the model does not explain the mechanisms of fouling, but only indicates the proportionality between the increase in hydrodynamic resistance and the mass of deposit on the membrane as filtration proceeds under some conditions. The values of $\hat{R}_c$ vary from 10$^{10}$ to 10$^{16}$ m/kg for different aquatic substances [22]. Babel et al. [23] found that $\hat{R}_c$ for a Chlorella algae culture changed from 10$^{11}$ to 10$^{12}$ m/kg when the growth condition became inhibitive. Foley [24] reviewed different factors affecting the permeability of the cake layer formed in dead-end microfiltration of microbial suspensions. It was found that $\hat{R}_c$ is dependent on cell morphology, surface properties, operating pressure, and time. The resistance-in-series model has been used frequently to analyze membrane fouling phenomenon. Although it is easy to apply, one should be cautious in the use of this model as it doesn't consider pore blocking mechanism.

Kosvintsev et al. [25] developed another model to describe fouling by physical sieving of low pressure membranes by particles larger than membrane pore sizes. According to their analysis, membrane fouling by cake filtration does not start right after the onset of filtration, and the fouling is rather dominated by pore blocking until the membrane surface is covered by particles. This model describes the permeate volume as a function of permeate time, dominated by pore blocking at constant pressure as follows:

$$V = \frac{1}{\gamma n \beta} ln\ (1\ + \beta t^*) \tag{3}$$

where $V$ is the permeate volume (cm³), $\beta$it is the ratio of the membrane area fouled with particles to the area of clean pores. This constant must be identified from experimental measurement for a given membrane and it should be slightly greater than unity. $n$ the number of particles per unit volume of the feed, $\gamma$ is the ratio of the pores area to the total membrane area and $t^*$ is the dimensionless filtration time $= \gamma \, n \int_0^t \frac{dV}{dt}$. More details of the model are presented in the authors' recent work [26]. This model was limited to pore blocking fouling and the effect of cake layer on the permeate volume was not considered.

Zydney et al. combined two fouling mechanisms, pore blockage and cake formation, to describe fouling of low pressure membranes by proteins and humic acids [27, 28]. Again, this model is established by assuming that the fouling is caused primarily by large particles, aggregates of proteins and humic acids. The mathematical development is based on constant pressure operation and varying flux, and it can be written as below:

$$\frac{J}{J_o} = \exp\left(-\frac{K_b \Delta P C_b}{\mu R_m} t\right) + \frac{R_m}{R_m + R_c}\left[1 - \exp\left(-\frac{K_b \Delta P C_b}{\mu R_m} t\right)\right] \tag{4}$$

where J and $J_0$ (m³/s) are the permeate flux at a given time and the initial flux through the unfouled membrane respectively, $K_b$ (m²/kg), a pore blockage parameter, is equal to the blocked membrane area per unit mass of aggregates convected to the membrane. This parameter can be measured experimentally. $C_b$ (kg/m³) is the bulk concentration of large aggregates, $R_m$ (1/m) is the clean membrane resistance, Rc is cake layer resistance (1/m), $\mu$ is the solution viscosity and $\Delta P$ is the transmembrane pressure (Pa). Both resistances can be measured experimentally. The right-hand side of the equation has two terms that are related to pore blocking and cake formation, respectively. The first term (pore blocking) dominates the early stage of fouling, and the second term (cake filtration) governs fouling at longer times. The impact of solution chemistry on membrane fouling is, however, not included in the model, but was rather considered as a prerequisite for the aggregation of proteins or humic acids.

In comparison to the aforementioned models, adsorptive fouling of membranes by particles smaller than membrane pore sizes is incorporated in the following model. The impact of the adsorption layer on the permeability of membranes can be estimated using a modified form of Hagen-Poiseulle capillary filtration model [29] as below:

$$\frac{J}{J_o} = [1 - \frac{\delta'}{r_p}]^4 \tag{5}$$

where J and $J_0$ (m³/m².s) are the permeate flux after the formation of the adsorptive fouling layer and the initial flux, respectively, under a given transmembrane pressure, $\delta'$ (m) is the thickness of the adsorption layer that can be measured experimentally and $r_p$ (m) is the membrane pore radius. The major difficulty in applying the adsorptive fouling model to filtration of natural surface waters lies in the complex nature of aquatic NOM. In other words, the value of $\delta'$ is not easy to obtain either theoretically or experimentally. This problem is further complicated by the heterogeneity of membrane surface properties.

# 4. Chemical attachment of foulants on membrane surfaces

An underlying question on membrane fouling is the origin of the attachment of foulants on the membrane surface. The major forces contribute to attachment are dispersion interaction force and polar interactions force [30]. These forces apply to material entities at different scales.

## 4.1. Chemical attachment by dispersion interaction

Foulants stay together on membrane surfaces most likely due to the presence of physiochemical interactions, such as the dispersion interaction between aqueous entities. This dispersion interaction is due to Van der Waals attractive force between molecules across water and is balanced by the electrostatic repulsion between particles and the membrane surface due to the presence of surface charges. As shown in energy curve figure (2) the height of the energy barrier depends not only on how strong the attractive interaction is, but also on the magnitude of the repulsive electrostatic interaction. Therefore, it is usually considered beneficial to increase the charge density of the similarly charged interacting entities to reduce attachment.

To represent the dispersion interaction, the Hamaker constant can be used. It is the property of a material, which represents the strength of van der Waals interactions forces between macroscopic bodies through a third medium as shown in Figure (3). Typical values of the Hamaker constant are in the range of $10^{-19}$ - $10^{-21}$ Joules. It can be estimated using the Lifshitz theory of macroscopic van der Waals interactions forces, which ignores the atomic structures of the interacting molecules, and calculates the forces between them in terms of their dielectric constants ($\varepsilon$) and refractive indices (n) [31, 32].The Hamaker constant, A, for two macroscopic phases 1 and 2 interacting across a medium 3 is approximated as:

$$A \approx \frac{3}{4} KT \frac{\varepsilon_1 - \varepsilon_3}{\varepsilon_1 + \varepsilon_3} \cdot \frac{\varepsilon_2 - \varepsilon_3}{\varepsilon_2 + \varepsilon_3} \cdot \frac{3 h v_e}{8\sqrt{2}} \cdot \frac{\left(n_1^2 - n_3^2\right)\left(n_2^2 - n_3^2\right)}{\left(n_1^2 + n_3^2\right)^{1/2} \left(n_2^2 + n_3^2\right)^{1/2} \left\{\left(n_1^2 + n_3^2\right)^{1/2} + \left(n_2^2 + n_3^2\right)^{1/2}\right\}} \tag{6}$$

where "1" and "2" denote two interacting bodies inside medium "3", A is the Hamaker constant, $ve$ is the medium absorption frequency (for $H_2O$, $ve = 3 \times 10^{15}$ $s^{-1}$), $\varepsilon$ is the dielectric constant that indicates the extent to which a material concentrates electric flux, n is the refractive index, K is the Boltzamnn constant, h is the Plank constant and T is the absolute temperature [33].

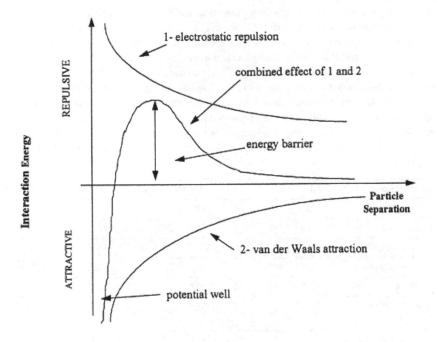

**Figure 2.** Energy curve of interaction forces [33]

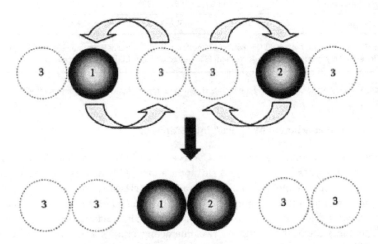

**Figure 3.** Interaction between 2 microscopic bodies 1 and 2 through medium 3 [33]

Table 1 lists the Hamaker constants representing the van der Waals interaction between polystyrene latex particles and different membrane materials across water, calculated using the macroscopic approach [30]. The Hamaker constant at zero frequency, Av=0, represents the static interaction and this term is always less than or closed to ¾ KT. The Hamaker constant at zero frequency is less than the total strength of van der Waals interactions forces. Hamaker constants at frequencies above zero, Av>0, is related to the three refractive indices, or fundamentally, the dispersion interaction between these surfaces. As shown in Table 1, the minimum and the maximum interaction force are observed in PTFE and alumina membranes with latex particles, respectively. The dispersion interaction between latex and PVDF is slightly less than half of that between two latex particles which indicates less irreversible fouling. [33]

| Interaction System [1] (1-3-2) | Dielectric Constant (kHz) [2] | | | Refractive Index [3] | | | Hamaker Constant x $10^{21}$ (J) | | |
|---|---|---|---|---|---|---|---|---|---|
| $\varepsilon 1$ | $\varepsilon 3$ | $\varepsilon 2$ | n1 | n3 | n2 | $A_{v=0}$ | $A_{v>0}$ | $A_{tot}$ | |
| Latex -Water- PTFE | 2.55 | 80 | 2.1 | 1.557 | 1.333 | 1.359 | 2.75 | 1.55 | 4.3 |
| Latex -Water- PVDF | 2.55 | 80 | 6.4 | 1.557 | 1.333 | 1.42 | 2.47 | 5.12 | 7.59 |
| Latex -Water- CA | 2.55 | 80 | 4.5 | 1.557 | 1.333 | 1.475 | 2.59 | 8.27 | 10.9 |
| Latex -Water- PP | 2.55 | 80 | 1.5 | 1.557 | 1.333 | 1.49 | 2.79 | 9.12 | 11.9 |
| Latex -Water- Cellulose nitrate | 2.55 | 80 | 6.4 | 1.557 | 1.333 | 1.51 | 2.47 | 10.2 | 12.7 |
| Latex -Water- PES | 2.55 | 80 | 3.5 | 1.557 | 1.333 | 1.55 | 2.65 | 12.5 | 15.1 |
| Latex -Water- Latex | 2.55 | 80 | 2.55 | 1.557 | 1.333 | 1.557 | 2.72 | 12.8 | 15.6 |
| Latex -Water- PC | 2.55 | 80 | 2.95 | 1.557 | 1.333 | 1.586 | 2.69 | 14.4 | 17.1 |
| Latex -Water- Alumina | 2.55 | 80 | 11.6 | 1.557 | 1.333 | 1.75 | 2.16 | 22.9 | 25.1 |
| Latex -Water- fused quartz | 2.55 | 80 | 3.8 | 1.557 | 1.333 | 1.448 | 2.63 | 6.74 | 9.37 |

Note: 1 PTFE: Polytetrafluoroethylene, PVDF: Polyvinylidene fluoride, CA: Cellulose acetate, PP: Polypropylene, PES: Polyethersulfone, PC: Polycarbonate; dielectric constant [31], Refractive index [32]; $A_{tot} = A_{v>0} + A_{v=0}$

**Table 1.** Hamaker constants calculated using the Lifshitz equation for representative particle-membrane interaction systems [30]

## 4.2. Physiochemical attachment by "polar" interactions

The Derjaguin and Landau, Verwey and Overbeek ( DLVO) theory has been extended, including different types of interactions, to applications with aqueous phase. Van Oss postulated the concepts of apolar and polar interactions to classify and predict these interactions [30]. The apolar interaction mainly consists of dispersion interaction. On the other hand, the polar (or Lewis acid-base) interaction is comprised of the interactions between Lewis acid-base pairs in the system, including the two interacting entities and surrounding water molecules. These interactions are useful in explaining the advantage of hydrophilizing the membrane surface to reduce the irreversible attachment of particles and other fouling materials on membrane surface. According to the concept of apolar/polar interactions, the strength of chemical attachment depends not only on the dispersion interaction (apolar), but also, or even more dominantly, on the polar interactions. The latter can be either attractive or repulsive based on the hydrophilicity of the two interacting surfaces.

For two hydrophilic surfaces, the polar interaction is repulsive and counteracts the attractive dispersion interaction. Therefore, the total interaction becomes either weakly attractive or repulsive even in the absence of electrostatic repulsion which leads to reduce fouling. In comparison, the polar interactions would be fairly attractive between hydrophobic surfaces, which are additive to the attractive dispersion interactions. Consequently, electrostatic repulsion becomes the dominant factor in balancing the attractive and repulsive interaction which enhances fouling. Therefore, there are in principle at least two possible approaches to make the membrane less vulnerable to the attachment of aquatic contaminants: hydrophilization of membrane surfaces (to enhance thermodynamic stability) and ionization of membrane surfaces (to achieve kinetic stability). Both approaches have been investigated by several researchers [34, 35-39]. The presence of polar interaction has also been used to explain different affinities of silica and latex particles on hydrophilic membranes [40]. Hydrophobic polystyrene latex particles showed less affinity to three commercial hydrophilic membranes than silica particles, as measured using atomic force microscopy (AFM). The hydrogen bonding attraction between silica particles and membrane surfaces was speculated to be the primary reason for the greater attachment. Regardless of the true mechanisms, such results suggest that the molecular structure of membranes and aquatic particles can be important to their interactions. Another complicated problem is the presence of NOM in natural water. The sorption or deposition of NOM moieties on particle and membrane surfaces can form an additional polymeric layer at solid/water interfaces.

### 4.3. Chemical attachments between heterogeneous surfaces

All previous chemical attachment mechanisms are based on the assumption that the interacting surfaces have homogeneous surface properties, and thus can be characterized using some global parameters, such as: charge density, hydrophobicity, and the Hamaker constant. However, this may not be realistic because particles could have heterogeneous surfaces. Different parts of the surface have different affinities to the membrane. In addition, the membrane surface, especially that modified, also likely has heterogeneous surface properties relevant to foulant attachment. This heterogeneity can be attributed to different physical and/ or chemical origins. For instance, the attachment of particles to membrane pores of various shapes was investigated. It was found that membrane pores with round corners are the least affinitive to colloidal fouling compared to those with sharp and spiky corners due to enhanced electrostatic repulsion [41]. In another investigation, the surface heterogeneity of nanofiltration and reverse osmosis membranes was studied using chemical force microscopy, a modified technique based on AFM to obtain the lateral distribution of surface energies/stickiness. It was found that the surfaces of the two membranes used were chemically heterogeneous, and that the heterogeneity became more significant below micron-sized dimensions [42]. This implies that the stickiness of membrane surfaces to foulants can be heterogeneous, rather than uniformly homogenous as considered previously.

# 5. Factors affecting fouling

- Membrane properties: pore size, hydrophobicity, pore size distribution and membrane material.
- Solution properties: solid (particle) concentration, particle size and nature of components.
- Operating conditions: pH, temperature, flow rate and pressure.

## 5.1. Membrane properties

In an aqueous environment a membrane can be attractive or repulsive to water. The composition of the membrane and its corresponding surface chemistry determine its interaction with water, thus affecting its wettability. The wettability of the membrane can be determined by measuring the contact angle between the membrane surface and a droplet of liquid, as shown in Figure (4). Hydrophilic membranes are characterized by the presence of active groups that have the ability to form hydrogen-bonds with water and so these membranes have wettability as can be seen in Figure (4.b). Hydrophobic membranes have the opposite interaction to water compared to hydrophilic membranes as they have little or no tendency to adsorb water and water tends to bead on their surfaces (i.e. discrete droplets) as shown in Figure (4.a). This tends to enhance fouling. Hydrophobic membranes possess low wettability due to the lack of active groups in their surface for the formation of hydrogen-bonds with water. Particles, which foul membranes in aqueous media, tend to be hydrophobic. They tend to cluster or group together to form colloidal particles because this process lowers the interfacial free energy. Usually, greater charge density on a membrane surface is associated with greater membrane hydrophilicity. Polysulfone, cellulose acetate, ceramic and thin-film composite membranes used for water treatment and wastewater recovery typically carry some degree of negative surface charge and hydrophilic. Thus, fouling can be reduced with use of membranes with surface chemistry which have been modified to render them hydrophilic.

(a)                                         (b)

**Figure 4.** a) Hydrophobic membrane, (b) Hydrophilic membrane [14]

Membrane morphology also has a considerable effect on fouling as pore size, pore size distribution and pore geometry especially at the surface of the membrane. These determines the predominant fouling mechanisms such as pore blocking and cake formation as previously discussed in section 2.

## 5.2. Solution properties

The properties of the feed solution also significantly influence membrane fouling. Some of the important feed properties are solid (particle) concentration, particle properties, pH and ionic strength. Generally, an increase in the feed concentration results in a decline in the permeate flux. This is due to the increase in membrane fouling by the presence of a higher foulant concentration. Particles may be present in the feed because of the nature of the feed or through precipitation of soluble feed component(s). The particles can cause fouling by pore blocking, pore narrowing or cake formation, dependent on the particle sizes. Higher permeate fluxes and cake thicknesses are usually obtained with larger particles [43]. Large particle size is one of the factors that inhibit deposition. In a filtration process, the particle sizes in the feed often cover a wide range. The presence of fine as well as coarse particles results in a lower cake porosity as the fine particles can slide between the large ones, filling the interstices. The range of the particle size distribution plays a major role in the selective deposition at high crossflow. In addition to the particle size, the particle shape affects the porosity of the cake formed on the membrane surface. In general, the lower the particle sphericity, the greater is the porosity [43].

Some other factors, such as: pH, ionic strength, and electric charges of particles, are also important. The pH and ionic strength of the feed affect the charge on the membrane, the charge on the particles, conformation and stability of, and thereby adhesiveness of particles/molecules and the size of the cake. For example, a study of the impact of pH of the latex emulsion on membrane fouling showed that the latex emulsion pH should be high enough to prevent the coagulation of latex particles, and hence, to increase the antifouling properties of the latex emulsion. Also, it has been showed that a reduction in pH could decrease the molecular size of NOM and thus enhances adsorption onto membrane, resulting in a significant fouling.

## 5.3. Operating conditions

The effect of temperature on the permeate flux was investigated and found that at higher temperatures, the permeate flux increased, indicating a lower degree of fouling. Changing the feed temperature from 20°C to 40°C lead to an increase in the permeate flux up to 60% [44]. This might be due to the fact that changes in the feed water temperature resulted in changes in the permeate diffusion rate through the membrane.

The cross-flow velocity is defined as the superficial velocity of the feed stream travelling parallel to the membrane surface. The effect of the cross-flow velocity on permeate flux has been studied for a wide variety of feed solutions. It is believed that increasing the cross-flow velocity positively affects the mass transfer coefficient of the solute and the extent of mixing near the membrane surface [45]. Consequently, the permeate flux is increased with cross-flow velocity. Higher mixing experienced with larger cross-flow velocity leads to a reduction of aggregation of the feed solids in the gel layer, essentially due to increasing diffusion of these components back towards the bulk, leading to an overall reduction in the effect of concentration polarization.

The control of the transmembrane pressure (TMP) which is the pressure difference between the feed and permeate stream is essential as it greatly affects the permeation rate. At a higher TMP, the force of the fluid flowing towards the membrane is increased, leading to a higher permeate flux. Increasing the applied pressure influence the permeate flux as illustrated in Figure (5). At very low pressure $p_1$, the flux is close to pure water flux at the same pressure. As the applied pressure is increased to pressure $p_2$, the higher flux causes increased concentration polarization of the retained material at the membrane surface increases. If the pressure is increased further to $p_3$ which considered the critical pressure, concentration polarization becomes enough for the retained solutes at the membrane surface to reach the gel concentration, $c_{gel}$. Once a gel layer has formed, further increase in the applied pressure does not lead to further increase in the permeate flux above this critical value. The gel layer thickness and the density of the retained material at the membrane surface layer, however, increase. This limits the permeate flux through the membrane, and hence, the flux reaches a steady state level. It was reported that no fouling was experimentally observed when the process was operated under this critical flux [14].

# 6. Remediation of membrane fouling

Fouling remediation can be done through pre-treatment the feed to limit its fouling propensity, improving the antifouling properties of the membrane, membrane cleaning and backwash conditions and optimization of the operating conditions already discussed previously.

## 6.1. Feed pre-treatment

Membranes are susceptible to fouling; therefore, pretreatment of the feed is required to control colloidal, organic, and biological fouling as well as scaling. The pretreatment scheme must be capable of controlling membrane fouling to such an extent that a practical cleaning frequency can be achieved. For low-pressure membranes, a number of pretreatment methods are currently used.

### 6.1.1. Coagulation

Coagulation involves the addition of chemicals coagulants, such as: $FeCl_3$, $FeSO_4$, alum, polyaluminum chloride, etc.., to increase the size of suspended and colloidal particles in the feed prior to filtration. It was found that reversible fouling was reduced with coagulation pretreatment, but the extent of irreversible fouling was unchanged. This can be attributed to the fact that large particles are formed from small particles, and hence, reversible fouling decreases with the use of coagulation. However, smaller particles, which are not coagulated, still remain in the feed and causes irreversible fouling. Factors affecting membrane fouling includes coagulant dosage, pH, nature of dissolved organic matters as well as $Ca^{2+}$ content of the feed water [14]. Moreover it was found that following coagulation pretreatment, most membrane

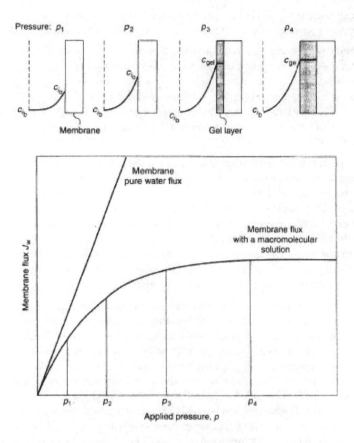

**Figure 5.** The effect of pressure on membrane flux [14]

fouling was due to the smaller hydrophilic NOM particles [13]. This finding is consistent with the fact that most metal-based coagulants are known to preferentially remove hydrophobic rather than hydrophilic substances. Coagulation reduced the rate of membrane fouling by minimizing pore plugging and increasing the efficiency of membrane backwashing.

Coagulation can be done by In-line coagulation process (IC), which refers to the dosing of coagulant into the feed stream. Rapid mixing in the feed stream allows the flocs to form (but not to settle) and finally enter the filtration unit (e.g., UF). Therefore, In-line coagulation doesn't require the sedimentation or prefiltration step prior to UF. Despite a larger fouling load in terms of suspended matter, IC may improve membrane performance due to the change in the fouling mechanism to cake formation rather than pore blocking. Once a cake is built up, it can be removed by backwashing easily. For in-line coagulation, the influence of membrane polymer nature and structure on fouling is alleviated. Cleaning frequency is also reduced and cleaning aggressiveness could be lowered. Consequently, the permeate flux increases, and the effect of seasonal water quality variations on filtration can be better controlled [46].

Sedimentation process can be used following the coagulation process. In this combined pretreatment method, a coagulant is applied and the formed flocs are settled out by sedimentation. The supernatant is then fed to the membrane filtration unit. In one study at East St. Louis, when UF was used after coagulation-sedimentation (CS) for 400 h, no membrane fouling was observed [47]. The coagulation or CS pretreatment process was very effective in increasing UF membrane life because this process removed the primary foulants such as high molecular weight humics [48].

Alternative process is coagulation-adsorption, which refers to adsorption of foulants using an adsorbent such as powdered activated carbon (PAC) between the coagulation step and UF. In one study, wastewater with the initial COD of 165 mg/L and turbidity of 90 NTU was treated with 120 mg $FeCl_3$/L at pH of 5.5. The COD and turbidity of the treated water were reduced to 23 mg/L and 12 NTU, respectively. When a further treatment step by adsorption with PAC was used, the COD dropped further to 7 mg/L [49]. The use of adsorption (PAC) and coagulation (alum and polyaluminum chloride) as pretreatment steps prior to membrane filtration was also investigated to remove organics. Significant improvement in the removal of organic materials and trihalomethane precursors were obtained [50].

Flocculation is another pretreatment method that can remove particles and colloids and hence improve the permeate flux. It is used to achieve three objectives: eliminating the penetration of colloidal particles into the membrane pores, increasing the critical flux, and modifying the characteristics of the deposits. The use of flocculation prior to membrane filtration reduced clogging of the membrane by aggregating smaller particles, thereby retaining them on the surface of the membrane. The larger flocs on the membrane surface are washed off by the retentate due to the tangential force (cross-flow) of the incoming solution, thus preventing membrane clogging. Flocculation can be used in combination with coagulation. Flocculation enhances the formation of larger flocs from particle aggregates generated by coagulation. In addition, flocculants induce floc formation from smaller particles that would not form particle clusters by coagulants.

### 6.1.2. Magnetic ion exchange

Magnetic ion exchange (MIEX) is a chemical process in which dissolved ions and charged species in water are adsorbed to polymer beads. Once they are saturated, the beads can be recovered and regenerated using a brine solution to desorb the charged species and ions. As a large percentage of the dissolved organic carbon (DOC) is polar, so it can be removed by MIEX, by exchanging chloride ions on the resin surface for polar dissolved and colloidal organic materials. Numerous studies have shown that ion exchange preferentially removes high charge density, medium-to-low molecular weight organic materials, which can consist of hydrophobic, transphilic and hydrophilic organic fractions. Ion exchange can therefore be synergistic with coagulation in reducing DOC loading entering the membrane unit, where coagulation removes the lower charge density, higher molecular weight hydrophobic fractions. A number of DOC removal methods were compared: alum coagulation (without pH control), alum coagulation (with pH controlled at 6), ion exchange using MIEX resin, and combined treatment of alum coagulation and MIEX. The relative effectiveness of those pretreatment methods for DOC removal was ranked in the order: alum/ MIEX > MIEX > alum pH 6 > alum (no pH control) [51]. Also, it was found that MIEX could remove more NOM than coagulation process could, even at very high coagulant concentrations [52]. When it is used as a pretreatment step, up to 80% of NOM can be removed prior to UF. Moreover, combining coagulation with MIEX was found to be able to remove 90 % of trihalomethane and haloacetic acid precursors from water [53].

### 6.1.3 Micellar-Enhanced filtration

Micellar enhanced ultrafiltration is an emerging technique that it is used to improve the performance of a filtration process by adding a surfactant to the feed in order to promote the entrapment of foulants in the micelles formed by the surfactant. Surfactants are molecules that contain a hydrophobic tail (usually long chain hydrocarbon) and a hydrophilic head. Above a specific concentration, surfactant molecules come together to form clusters or micelles. This concentration is termed the critical micelle concentration (CMC) and differs depending on the type of surfactant.

There are numerous types of surfactants used in industry today, categorized by the charge of the hydrophilic portion of the molecule: anionic (negatively charged), cationic (positively charged), non anionic (neither positively nor negatively charged), and zwitterionic (both negatively and negatively charged). The formation of micelles increases the particle size, allowing the use of membranes with larger pore sizes for the same feed. Some surfactants also interfere with hydrophobic interactions between bacteria and membranes. In addition, surfactants can disrupt functions of bacterial cell walls. Therefore, they reduce fouling dominated by the biofilm formation. The choice of a surfactant is based on its compatibility with the solid for the solid recovery and reuse and its effect on the filtration system. A study has been conducted to compare the use of dodecylbenzesulfonic acid, as an anionic surfactant, and dodecylamine, as a cationic surfactant, to improve the removal of lead and arsenic from municipal wastewater [54]. It was concluded that while both surfactants enhanced separation of the heavy metals, the cationic surfactant was more effective than the anionic one. In another study, sodium dodecyl sulphate (SDS), as an anionic surfactant, and trimethylammonium bromide (CTAB), as cationic surfactant, were used to improve ultrafiltration of latex paint wastewater [55]. With SDS at twice its CMC, a reduction of 58% of permeate flux was observed. In contrast, using CTAB at twice its CMC, the permeate flux increased up to 134%. The effectiveness of surfactant also depends on the membrane material and its surface charge. One study indicated that for hydrophilic membranes, the permeate flux was reduced when ethoxylated alkyl phenol alcohol (Triton X-100), a non-ionic surfactant, was used above its CMC. However, for hydrophobic membranes, no significant flux reduction was observed with the same surfactant [56].

## 6.2. Membrane properties modification

Membrane properties affect the solute-membrane interaction and, consequently, the extent of adsorption and fouling. For filtration of proteins, since proteins adsorb more strongly to hydrophobic surfaces than hydrophilic ones, the use of hydrophilic membranes (cellulose esters, aliphatic polyamides) can help reducing membrane fouling. Chemical modification of a membrane (for example, sulfonation of polysulfone) or blending a hydrophobic polymer (polyetherimide, polyvinylidenefluoride) with a hydrophilic one (polyvinylpyrrolidone) can enhance the anti-fouling property of membranes. Another way to influence the solute-membrane interaction can be achieved by the pretreatment of the membrane with hydrophilic surfactants or enzymes. Conventional ultrafiltration membranes, such as: polysulfone, polyethersulfone or polyvinylidene fluoride, can be made more hydrophilic by surface modification using various methods [57, 58]:

- Plasma treatment of the membrane surface;

- Polymerization or grafting of the membrane surface initiated by UV, heat or chemicals;

- Interfacial polymerization;

- Introduction of polar or ionic groups to the membrane surface by reaction with bromine, fluorine, strong bases and strong acids.

Hydrophilization of the membrane surface also can be done by pre-coating the membrane with a nonionic surfactant. This method is very attractive for practical application because of it is simple. With this treatment, ultrafiltration of antifoam rejection was improved significantly, and hence, the permeate flux was almost doubled [59]. Alternatively, ozone can be used to modify the membrane surface and its hydrophobicity. This treatment introduces peroxide groups to the polymer surface, which can initiate graft polymerization of monomers with hydrophilic groups, and thus improves the hydrophilicity of the polymer surface. The concentration of peroxide groups formed can be used to determine the effectiveness of the ozonation process. The effect of ozonation on the permeate flux was studied using a polysulfone UF membrane. It was found that ozonation increased the permeate flux by 10%, and membrane surface oxidation by the mixture of ozone and hydrogen peroxide was even more effective. Ozone prolonged the period required to reach appreciable fouling rather than eliminated it [57]. The applied ozone dose and ozonation time determine the amount of peroxide groups generated and thus the degree of hydrophilicity enhancement of the membrane surface.

## 6.3. Membrane fouling cleaning

Membrane cleaning is an integral part of a membrane system operation and has a significant impact on the process operation. Fouling materials can be removed by hydraulic means such as backwashing or by chemical means such as enhanced backwash (EBW). Cleaning operation can be classified as cleaning in-place (CIP) or off-line chemical cleaning (or soaking). In CIP the membrane module is cleaned without removing it from the installation while in off-line cleaning the module is removed from the system and soaked in a chemical.

Backwashing is done by reversing the flow across the membrane, using the permeate to remove foulants accumulated on the membrane surface and/or clogged the membrane pores.

In EBW a cleaning chemical is added to the backwash water and the water is recirculated for a short period of time (10-15 min). Chemical cleaning is an integral part of a membrane process operation, which has a profound impact on the performance and economics of the process. Currently, types of cleaning chemicals used are recommended by membrane manufacturers. Some of them are proprietary cleaners while others are commercial chemicals. Chemical cleaning is required for the membrane to regain most of its permeability. Chemical cleaning is performed when flushing and/ or backwashing cannot restore the permeate flux. In chemical cleaning, the chemical dose is usual higher than that for the enhanced backwashing and the frequency of chemical cleaning is usual lower (approximately 1 per week). Moreover, the enhanced backwashing can be fully automated while the chemical cleaning involves manual labor due to its off-line operation. Proper selection of chemical cleaning agents, conditions for their application and understanding their performance are important. A cleaning agent is usually selected based on the types of foulants. The effectiveness of various operating strategies for different fouling types is summarized in Table 2. As indicated in Table 2, the chemical cleaning is an effective control strategy for a majority of membrane fouling types.

| Type of Fouling | Effects of Operating Strategy | | | |
|---|---|---|---|---|
| | Hydraulic Cleaning / Backwashing | Feed Chlorination | Feed Acidification | Chemical Cleaning |
| Inorganic | - | - | ++ | ++ |
| Particulate | ++ | - | - | ++ |
| Microbial | + | ++ | +* | ++ |
| Organic | - | + | - | ++ |

Notes: "-": No effect/ negative effects; "+": some positive affects; "++": positive effects; "*": together with feed chlorination.

**Table 2.** Effects of operating strategies on membrane fouling [14]

Calcium, magnesium and silica scaling, often a serious problem in reverse osmosis operation, is generally not a concern in ultrafiltration because these ions permeate through the membrane. Ultrafiltration of cheese whey, in which high calcium levels can lead to calcium scaling, is an exception. Because many feed waters contain small amount of soluble ferrous salts, hydrate iron oxide scaling is a problem. In ultrafiltration, these salts are oxidized to ferric iron by entrained air. Ferric iron is insoluble in water; hence, an insoluble iron hydroxide gel forms and accumulates on the membrane surface. Such deposits are usually removed with citric or hydrochloric acid wash. Chemicals commonly used for cleaning UF and MF membranes in water industry fall into five categories, as summarized in Table 3

| Category | Major Functions | Typical Chemicals |
|---|---|---|
| Caustic | Hydrolysis, solubilisation | NaOH |
| Oxidants / disinfectants | Oxidation, disinfection | $NaOCl$, $H_2O_2$, peroxyacetic acid |
| Acids | Solubilization | Citric, nitric, hydrochloric acid |
| Chelating agents | Chelation | Citric acid, EDTA |
| Surfactants | Emulsifying, dispersion, surface conditioning | Surfactants, detergents |

**Table 3.** Major categories of membrane cleaning chemicals [14]

Regardless of the membrane system used, chemical cleaning is cumbersome and requires shutdown of the unit. This results in a reduction of the overall plant capacity and produces a waste that may be difficult to dispose of. There are also concerns that repeated chemical cleaning might affect the membrane life. Chemical cleaning should thus be limited. Because membrane cleaning is essentially conducted through chemical reactions between cleaning chemicals and fouling materials, factors that affect the cleaning efficiency are concentration, temperature, length of the cleaning period and hydrodynamic conditions. The cleaning chemical concentration can affect both the equilibrium and the rate of the reaction. The cleaning chemical concentration plays a key role not only to maintain a reasonable reaction rate but also to overcome mass transfer barriers imposed by the fouling layer. In practice, the cleaning chemical concentration is usually high enough to ensure a desirable reaction rate. It is mass transfer, which dictates the limiting chemical concentration that is adequate for cleaning purpose.

Temperature can affect membrane cleaning by (1) changing the equilibrium of a chemical reaction, (2) changing the reaction kinetics, and (3) changing the solubility of fouling materials and/or reaction products during the cleaning. Generally, membrane cleaning is more efficient at elevated temperatures. However, compatibility of the membrane and other filter components regarding temperature should also be checked.

Membrane cleaning involves mass transfer of chemicals to the fouling layer and the reaction products back to the bulk liquid phase. Therefore, hydrodynamic conditions that promote contact between cleaning chemicals and fouling materials during cleaning are required. From a mass transfer point of view, dynamic cleaning involving circulating cleaning solutions through the system can be more effective than static cleaning such as soaking.

Moreover, mechanical cleaning can be used if chemical cleaning does not restore the permeate flux. Tubular membrane modules could be effectively cleaned by forcing sponge balls of a slightly larger diameter. The balls gently scrape the membrane surface, removing deposited materials. Sponge-ball cleaning is an effective but relatively time-consuming process, so it is performed rather infrequently.

# 7. Conclusion

Membrane fouling is a critical problem that reduces the permeate flux, requires periodic cleanings, and limits further membrane development due to the hindrance of wider application to various processes by fouling. Fouling is caused by the deposition of suspended or dissolved solids in the feed on the external membrane surface, on the membrane pores, or within the membrane pores. One of the two main factors, which has a significant effect on membrane fouling, is the membrane properties, such as: pore size and distribution, hydrophobicity and membrane material. Membranee fouling is a phenomenon involving the interaction between the membrane and the solution. Therefore, another important factor governing fouling is the solution properties, such as: concentration and nature of components and the particle size distribution. In addition, operational conditions such as pH, temperature, flow rate and pressure also greatly affect fouling.

Even though membrane fouling is inevitable during the filtration process, it can be controlled and alleviated. Current approaches to deal with membrane fouling include mathematical model prediction of membrane fouling and membrane fouling reduction using different techniques such as pre-treatment of the feed water, membrane modification, improving the operational conditions and cleaning. In order to determine the proper pre-treatment, a complete and accurate analysis of the feedwater should be made. In addition, the interaction of a particular membrane and a specific foulant needs be understood so that an appropriate method can be selected. Finally the fouling behaviour and autopsy protocol for membrane fouling can be concluded in four basic aspects: identification of fouling components, development of conceptual or phenomenological models for membrane fouling, establishment of mathematical models to describe or predict fouling, and development of fouling control strategies.

# Nomenclature

| Symbol | Unit | Physical Meaning |
|--------|------|------------------|
| $\beta$ | dimensionless | Ratio of membrane area influenced with the particles suspension to open area of the membrane pores |
| $C_b$ | kg/m³ | Bulk concentration of large aggregates |
| $J$ | m³/m².s | Permeate flux at any time |
| $J_0$ | m³/m².s | Initial permeate flux |
| $K_b$ | m²/kg | Pore blockage parameter |
| $m_d$ | kg/m² | Mass of deposits accumulated on unit surface area of membranes |
| $m_p$ | kg | Total mass of aggregates retained by the membrane |
| $\mu$ | Pa-s | Solution viscosity |

| Symbol | Unit | Physical Meaning |
|--------|------|------------------|
| n | dimensionless | Number of particles per unit volume |
| $\Delta P$ | Pa | Transmembrane pressure (TMP) |
| $r_p$ | m | Pore radius of membranes |
| $R_c$ | m$^{-1}$ | Hydrodynamic resistance of cake layers |
| $\hat{R}_c$ | m/kg | Specific resistance of cake layer on the membrane surface |
| $R_m$ | m$^{-1}$ | Hydrodynamic resistance of clean membranes |
| $t^*$ | dimensionless | Filtration time |
| $V$ | cm$^3$ | Permeate volume |
| $\delta'$ | m | Thickness of the adsorption layer |
| $\gamma$ | dimensionless | Ratio of the pores area to the total membrane area |
| $\varepsilon$ | dimensionless | Dielectric Constant |

## Author details

A. Abdelrasoul, H. Doan and A. Lohi

Department of Chemical Engineering, Ryerson University, Victoria Street, Toronto, Ontario, Canada

## References

[1] Hong, S., & Elimelech, M. (1997). Chemical and physical aspects of natural organic matter (NOM) fouling of NF membranes. *Journal of Membrane Science*, 132, 159-181.

[2] Quintanilla, V. A. Y., (2005). Colloidal and non-colloidal NOM fouling of ultrafiltration membranes: analysis of membrane fouling and cleaning. M. Sc. Thesis, UNESCO-IHE.

[3] Lee, S., Cho, J. and Elimelech, M. (2005). Combined influence of natural organic matter (NOM) and colloidal particles on nanofiltration membrane fouling. *Journal of Membrane Science*, 262, 27-41.

[4] Schafer, A.I. (2001). Natural organic matter removal using membranes, Ph.D. Thesis, UNESCO-IHE, UNSW, Australia.

[5] Mo, L., & Huanga, X. (2003). Fouling characteristics and cleaning strategies in a coagulation micro filtration combination process for water purification. *Desalination*, 159, 1-9.

[6]   Weisner, M. R., Clarke, M. M., Jacanglo, J.G., Lykins, B.W., Marinas, B. J., O'Mellia, C.R., Ritmann, B.E., and Semmens, M.J. (1992). Committee report: Membrane processes in portable water treatment. *Journal of the American Water Works Association*, 84(1), 59-67.

[7]   Aoustin, E., Schafer, A.I., Fane, A. G. and Waite, T. D. (2001). Ultrafiltration of natural organic matter. *Separation and Purification Technology*, 22-23, 63-78.

[8]   Makdissy, G., Croue , J.P., Buisson, H., Amy, G., and Legube, B. ,(2003). Organic matter fouling of ultrafiltration membranes. *Water Science and Technology Water Supply*, 3(5-6), 175-182.

[9]   Jucker, C., & Clark, M. M. (1994). Adsorption of aquatic humic substances on hydrophobic ultrafiltration membranes. *Journal of Membrane Science*, 97, 37-52.

[10]  Makdissy, G., Croue, J.P., Buisson, H., Amy, G., and Legube, B. (2003). Organic matter fouling of ultrafiltration membrane. *Water Science and Technology Water Supply*, 3(5-6), 175-182.

[11]  Kimura, K., Hane,Y., Watanabe , Y., Amy, G., and Ohkuma, N. (2004). Irreversible membrane fouling during ultrafiltration of surface water. *Water Research Journal*, 38, 3431-3441

[12]  Nilson, J., and Digiano, F. A. (1996). Influence of NOM composition on nanofiltration. *American Water Works Association*, 88, 53-66.

[13]  Carroll, T., King, S., Gray, S.R., Bolto, B. A., and Booker, N.A., (2000). The fouling of microfiltration membranes by NOM after coagulation treatment. *Water Research Journal*, 34, 2861-2868

[14]  Li, N.N., Fane, A.G., Winston, W. S. H., and Matsuura, T., (2008).Advanced membrane technology and applications, John Wiley& sons Inc.

[15]  Belfort, G.R.H., & Zydney, A.L.,(1994). The behavior of suspensions and macromolecular solutions in crossflow microfiltration. *Journal of Membrane Science*, 96, 1-58.

[16]  Javacek, M. H., & Bouchet, F. (1993) Constant flowrate blocking laws and an example of their application to dead-end microfiltration of protein solutions. *Journal of Membrane Science*, 82, 285-295.

[17]  Hermia, J. (1982). Constant pressure blocking filtration laws, application to power-law non-Newtonian fluids. *Transactions of the American Institute of Chemical Engineers*, 60, 183-187.

[18]  Granger, J., Leclerc, D., and Dodds, J.A. (1985). Filtration of dilute suspensions of latexes. *Filtration and Separation*, 22, 58-60.

[19]  Kim, A.S., & Hoek, E.M.V. (2002).Cake structure in dead-end membrane filtration: Monte Carlo simulations. *Environmental Engineering Science*, 19(6), 373-386.

[20] Zhang, M. & Song, L. (2000) .Mechanisms and parameters affecting flux decline in cross-flow microfiltration and ultrafiltration of colloids. *Environmental Science & Technology*, 34(17), 3767-3773.

[21] Chellam, S., Jacangelo, J.G., and Bonacquisti, T.P. (1998). Modeling and experimental verification of pilot-scale hollow fiber, direct flow microfiltration with periodic backwashing. *Environmental Science & Technology*, 32(1), 75-81.

[22] Endo, Y., & Alonso, M., (2001). Physical meaning of specific cake resistance and effects of cake properties in compressible cake filtration. *Filtration and Separation*, (9), 43-46.

[23] Babel, S., Takizawa, S., and Ozaki, H., (2002) .Factors affecting seasonal variation of membrane filtration resistance caused by Chlorella algae. *Water Research*, 36(5), 1193-1202.

[24] Foley, G., (2006). A review of factors affecting filter cake properties in dead-end microfiltration of microbial suspensions. *Journal of Membrane Science*, 274, 38-46.

[25] Kosvintsev, S., Holdich, R.G., Cumming, I.W., and Starov, V.M., (2002). Modelling of dead-end microfiltration with pore blocking and cake formation. *Journal of Membrane Science*, 208, 181-192.

[26] Kosvintsev, S., Cumming, I.W., Holdich, R.G., Lloyd, D., and Starov, V.M., (2004). Mechanism of microfiltration separation colloids and surfaces. *Physicochemical and Engineering Aspects*, 230, 167-182.

[27] Yuan, W., Kocic, A., and Zydney, A.L., (2002) .Analysis of humic acid fouling during microfiltration using a pore blockage-cake filtration model. *Journal of Membrane Science*, 198(1), 51-62.

[28] Ho, C.C., & Zydney, A.L., (2002) .Transmembrane pressure profiles during constant flux microfiltration of bovine serum albumin. *Journal of Membrane Science*, 209(2), 363-377.

[29] Srebnik, S., (2003). Polymer adsorption on multi component surfaces with relevance to membrane fouling. *Chemical Engineering Science*, 58(23-24), 5291-5298.

[30] Israelachvili, J., (1992). Intermolecular & surface forces. 2nd ed., San Diego, CA 92101, USA: Academic Press Inc. 450.

[31] Dielectric Constants of Materials ,http://www.clippercontrols.com/info/dielectric_constants.html

[32] Refractive index , http://www.texloc.com/closet/cl_refractiveindex.html

[33] John Gregory (2005), Particles in Water: Properties and Processes,Chapter 4

[34] Carroll, T., Booker, N.A., and Meier-Haack, J., (2002). Polyelectrolyte-grafted microfiltration membranes to control fouling by natural organic matter in drinking water. *Journal of Membrane Science*, 203(1-2), 3-13.

[35] Taniguchi, M., Kilduff, J.E., and Belfort, G., (2003) Low fouling synthetic membranes by UV-assisted graft polymerization: monomer selection to mitigate fouling by natural organic matter. *Journal of Membrane Science*, 222, 59-70.

[36] Hester, J.F., & Mayes, A.M., (2002). Design and performance of foul-resistant poly (vinylidene fluoride) membranes prepared in a single-step by surface segregation. *Journal of Membrane Science*, 202(1-2), 119-135.

[37] Wavhal, D.S., & Fisher, E.R., (2003) Membrane surface modification by plasma induced polymerization of acrylamide for improved surface properties and reduced protein fouling. *Langmuir*, 19(1), 79-85.

[38] Liu, Z.M., Xu, Z.K., Wang, J.Q., Yang, Q., Wu, J., and Seta, P. , (2003).Surface modification of microporous polypropylene membranes by the grafting of poly(gamma-stearyl- L-glutamate). *European Polymer Journal*, 39(12), 2291-2299.

[39] Yu, H.Y., Xie, Y.J., Hu, M.X., Wang, J.L., Wang, S.Y., and Xu, Z.-K. , (2005). Surface modification of polypropylene microporous membranes to improve its antifouling property in MBR: $CO_2$ plasma treatment. *Journal of Membrane Science*,254(1-2), 219-227.

[40] Brant, J.A., & Childress, A.E., (2004).Colloidal adhesion to hydrophilic membrane surfaces. *Journal of Membrane Science*, 241(2), 235-248.

[41] Bowen, W.R., & Sharif, A.O., (2002). Prediction of optimum membrane design: pore entrance shape and surface potential. colloids and surfaces. *Physicochemical and Engineering Aspects*, 201, 207-217.

[42] Brant, J.A., Johnson, K.M., and Childress, A.E., (2006). Characterizing NF and RO membrane surface heterogeneity using chemical force microscopy. colloids and surfaces. *Physicochemical and Engineering Aspects*, 280(1-3), 45-57.

[43] Vyas, H.K., Bennett, R.J., and Marshall, A.D., (2000). Influence of feed properties on membrane fouling in crossflow microfiltration of particulate suspension. *International Dairy Journal*, 10 ,855-861

[44] Salahi, A., Abbasi, M., and Mohammedi, T., (2010). Permeate flux decline during UF of oily wastewater. *Desalination*, 251,153-160.

[45] Salahi, A., Mohammedi, T., Pour, A., Rekabdar, F., (2000). Oily wastewater treatment using ultrafiltration. *Desalination*, 6,289-298.

[46] Doyen, W., (2003). Latest developments in ultrafiltration for large-scale drinking water. *Desalination, 113*, 165-177.

[47] Kruithof, J. C., Nederlof, M. M., Hoffman, J. A. M. H., and Taylor, J. S., (2004). Integrated membrane systems. *Research Foundation and American Water Works Association*, Elbert, Colorado.

[48] Minegishi, S., Jang, N.Y., Watanabe, Y., Hirata, S., and Ozawa, G., (2001). Fouling mechanism of hollow fibre ultrafiltration membrane with pre-treatment by coagulation/sedimentation process. *Water Science and Technology Water Supply*, 1(4), 49-56.

[49] Abdessemed, D., & Nezzal, G., (2002). Treatment of primary effluent by coagulation-adsorption-ultrafiltrationn for reuse. *Desalination*, 152, 367-373.

[50] Berube, P. R., Mavinic, D. S., Hall, E. R., Kenway, S.E., and Roett, K., (2002). Evaluation of adsorption and coagulation as membrane pretreatment steps for the removal of organic material and disinfection by product precursors. *Journal of Environmental Engineering and Science*, 1, 465-476.

[51] Drikas, M., Christopher, W., Chow, K., and Cook, D., (2003). The impact of recalcitrant on disinfection stability, trihalomethane formation and bacterial regrowth: A magnetic ion exchange resin (MIEX) and alum coagulation. *Journal Water Supply Research*, 52(7), 475-487.

[52] Slunjski, M., Bourke, M., and O'Leary, B., (2000). MIEX DOC process for removal of humics in water treatment. In Proceeding of IHSS – Australian Branch Symposium: Humic Substances Science and Commercial Applications. Monash University, Melbourne, 22-27.

[53] Singer,P.C., & Bilyk,K. (2002). Enhanced coagulation using a magnetic ion exchange resin, *Water Research* 36, 4009–4022.

[54] Ferella,F., Prisciandaro, M., Michelis, I., and Veglio, F., (2007). Removal of heavy metals by surfactant-enhanced ultrafiltration from wastewaters. *Desilination*, 207, 125-133.

[55] Bedasie,R., (2010).An investigation into the fouling phenomena of polycarbonate membranes used in the treatment of latex paint wastewater. M. Sc. Thesis.

[56] Byhlin, H. A., & Jonsson, A. S., (2002). Influence of adsorption and concentration polarisation on membrane performance during ultrafiltration of a non-ionic surfactant. *Desalination*,151 , 21-31

[57] Park, Y. G., (2002). Effect of ozonation for reducing membrane fouling in the UF membrane. *Desalination*, 147, 43-48.

[58] Mulder, M.H.V., (1993). Membranes in Bioprocessing, theory and application. Chapman and Hall, London, P.13.

[59] Noble, R.D., & Stern, S. A., (1995). Membrane separation technology principles and applications, Elsevier Science B.V., 46-83.

# Chapter 9

# Filtration of Radioactive Solutions in Jointy Layers

Mikhaylov Pavel Nikonovich,
Filippov Alexander Ivanovich and
Mikhaylov Aleksey Pavlovich

Additional information is available at the end of the chapter

## 1. Introduction

The solution of global energy problems of mankind, first and foremost associated with the development of nuclear energy. Already by 2030 the share of nuclear power generation in total electricity production should be about 25-30% (today - 16%). Currently, the total amount of radioactive waste in Russia is estimated at $5 \cdot 10^8$ m$^3$, the total β-activity of which is estimated at $7.3 \cdot 10^{19}$ Bq. At the same time on the liquid radioactive waste (LRW) accounts for about 85% of total activity, and their treatment and disposal become the most important task of nuclear energy.

One of the safest ways of disposal of waste of nuclear and chemical production is injection of them into deep-seated subterranean formations. Therefore, an important issue is to study the processes of the joint heat and mass transfer during the injection of waste into a porous collector layer to predict and control the state of the areas covered by the influence of radioactive impurities. The above forecast is carried out mainly by calculations, since the possibility of experimentally sizing of deep zones of contamination is very limited.

The processes of mass transfer in porous media have long been the object of study for many researchers. Have become classics of the G.I. Barenblatt [1], Bear J. [2, 3, 4], Bachmat Y. [5], A. A. Ilyushin [6], V. M. Keyes [7], L. D. Landau [8], R. I. Nigmatullin [9], V. N. Nikolayevsky [10, 11], L. I. Sedov [12]. In the works of Prakash A. [13], A. A. Barmin [14], E. A. Bondarev [15], M. L. Zhemzhurov [16, 17], E. V. Venetsianov, R. N. Rubinstein [18] the problems of the filtration of solutions, taking into account the phenomenon of adsorption, are regarded. Fluid flow through porous materials [19-24] are coincided to be well studied. Study of models of multi-component flows is devoted to the work of R. E. Swing [25].

Problems of disposal of radioactive waste in geological formations and the resulting ecological problems discussed in works of A. S. Belitsky, E. Orlova [26], A. Rybalchenko, M. K. Pimenov [27]. Modeling of temperature and radiation fields examined in works of D. M. Noskov, A. D. Istomin, A. G. Kessler, A. Zhiganov [28-29] (Seversk Technological Institute), I. Kosareva, and E. V. Zakharova [30-31] (Institute of Physical Chemistry RAS), and other researchers. In the works of A. Lehova, Y. Shvarova studied the rate of radionuclides in groundwater, the behavior of radioactive waste in the earth's crust after the injection. At the same time remain relevant problem of determining the concentration dependence of the fields on the parameters of injection of radioactive impurities, injection technology on the parameters of layers, etc.

The study of filtration processes in multilayer formations, as well as any thermodynamic problems of contacts of the bodies and environments, leads to the necessity of solving the problems of conjugation. To solve these problems are widely used numerical methods. The analytical solutions are constructed only for simple cases, such as linear flow in mass-isolated formation [32, 33]. And as the disposal at the request of the IAEA carried out on the timing of the order of tens thousand years, in these circumstances, the porous layer, can hardly be considered mass-isolated.

In this paper, in example of study of the filtration process of radioactive solutions, represented a modification of the asymptotic method, allowing successfully construct approximated solutions to conjugacy problems.

## 2. The mathematical formulation of the problem of heat and mass transfer in fluid flow with radioactive contaminant in the deep layers

Let us consider problem of heat and mass transfer, which describe the interrelated fields of concentration and temperature of the radioactive contaminant in the porous layer, through which flows a liquid with impurities, and the covering and the underlying layers are water-proof.

Typically, in deep horizons an aqueous solution is injected. This solution consist of a different soluble chemical compounds formed during the acid treatment process of structural elements of reactors and other parts of the design (process waste), or in the decontamination of buildings, cars, clothing and so on (non-technological waste) and includes a mixture of various radioactive nuclides [34]. Quite naturally the initial density of the solution divided into two factions

$$\rho_{tot} = \sum_{k=0}^{Nch} \rho_k + \sum_{i=1}^{Nrc} \rho_i, \tag{1}$$

where $\rho_k$ is the density of the dissolved non-radioactive components (for $k = 0$, we obtain the density of the solvent) $\rho_i$ is the concentration of radioactive $i$-th nuclide, $N_{ch}$, $N_{rc}$ - the number of different non-radioactive and radioactive components in the solution, respectively.

Consider an arbitrary reaction volume $dV$ in a porous layer containing a multi-component mixture (1). Mass flow passing through the surface of $dS$ reaction volume $dV$ can be represented as the sum of four terms

$$\frac{\partial}{\partial \tau}\int \rho_j dV = \int_S \vec{j}_j \vec{n} dS + \Delta_j^{ch} + \Delta_j^{reac} + \Delta_j^{ex},\qquad (2)$$

where the first term takes into account the mass exchange with the environment through the diffusion and convection currents, the second term describes the rate of change of mass in chemical reactions, the third term takes into account the change in mass due to radioactive decay of radionuclides and the fourth term describes the mass transfer processes between the components of the solution and formation.

Denote by $p$ the number of chemical reactions involving a $j$-component, and $\omega_i$ is the reaction rate per unit reactor volume, while the second term on the right side of (2) can be written as

$$\Delta_j^{ch} = \sum_{i=1}^{p} k_{ji} \int \omega_i dV,\qquad (3)$$

where $k_{ji}$ - the stoichiometric coefficients of chemical reactions of $j$-component.

The third term can be represented as

$$\Delta_j^{reac} = -\delta \int (\alpha \rho)_j dV,\qquad (4)$$

where $\alpha_j$ - the radioactive decay constant of $j$ -th radionuclide, $\delta$ - Kronecker delta function, equal to either one if the of $j$-component of the radionuclide, or zero if otherwise.

We assume that the transition of the impurity molecules of the liquid in the skeleton and its transition from a skeleton into a liquid are determined by the chemical potentials $\mu_s$, $\mu_w$. The fourth term is of the form

$$\Delta_j^{ex} = \int g(\mu_w, \mu_s) dV - \int g(\mu_j, \mu_s) dV,\qquad (5)$$

where $g(\mu_j, \mu_s)$ - a function of mass transfer between the of $j$-component of the solution and the skeleton of rock, $g(\mu_w, \mu_s)$ - mass transfer function corresponding to the transition of matter from the rock matrix in the solution.

Substituting (3) - (5) into (2) and transforming the surface integral into a volume integral, we obtain

$$\frac{\partial}{\partial \tau}\int \rho_j dV = -\int \mathrm{div}\left(\vec{j}_j\right)dV + \sum_{i=1}^{p} k_{ji}\int \omega_i dV - \delta\int (\alpha \rho)_j dV + \int g\left(\mu_w, \mu_s\right)dV - \int g\left(\mu_j, \mu_s\right)dV . \qquad (6)$$

By the arbitrariness of the reaction volume $dV$ and the continuity of the functions under the integral, we obtain

$$\frac{\partial \rho_j}{\partial \tau} = -\mathrm{div}\left(\vec{j}_j\right) + \sum_{i=1}^{p} k_{ji}\omega_i - \delta(\alpha \rho)_j + g\left(\mu_w, \mu_s\right) - g\left(\mu_j, \mu_s\right). \qquad (7)$$

The resulting equation is nonlinear, even in simple cases the values $\omega_i$ are polynomial functions of concentration. Therefore, in general, equation (7) forms a system of nonlinear partial differential equations. The solution of this system is quite complicated both mathematically, and in terms of its applicability to the description of particular phenomena.

Let us estimate in (7) the contribution of the second term. Obviously, the maximum change of mass in chemical reactions, while other things being equal, will be observed in the following two cases:

$A + B \rightarrow C \uparrow$ (evaporation),
$A + B \rightarrow C \downarrow$ (precipitation).

In both reactions the dissolved substances are excluded from consideration, which entails a decrease in the concentrations of the components of the solution. But this type of unpredictable chemical reactions creates the conditions for dangerous situations and in the deep burial of radioactive waste should be excluded. The chemical reaction scheme (acid-base and redox)

$$A + B + ... \rightarrow A' + B' + ...\left(\pm \Delta H^0\right). \qquad (8)$$

are valid and give a slight variation in the concentration of the solution, because typical of enthalpy $\Delta H^0$ change is of the order of several hundred kilojoules per mole of interacting substances and the corresponding change in mass $10^{-7} \div 10^{-12}$ kg, which is negligible in comparison with the mass of dissolved chemical components. Therefore, the change in mass due to chemical reactions will be neglected.

As shown in [35], the time of mass transfer between the fluid and the skeleton of the order of 0.1 s. Thus, the mass transfer, which is characterized by a concentration gradient, is almost instantaneous compared to the time of injection of pollutant that may be from several months to several years. Let us also neglect the processes of chemical compounds leaching from the porous rock to the solution, i.e. assume the condition $g(\mu_w, \mu_s) - g(\mu_j, \mu_s) \approx 0$.

Based on the above, equation (7) takes the form

$$\frac{\partial \rho_j}{\partial \tau} = -\mathrm{div}\left(\vec{j}_j\right) - \delta\left(\alpha \rho\right)_j.$$  (9)

Divide the resulting equation into two components: non-radioactive and radioactive fractions

$$\frac{\partial \rho_k}{\partial \tau} + \mathrm{div}\left(\vec{j}_k\right) = 0, \frac{\partial \rho_i}{\partial \tau} + \mathrm{div}\left(\vec{j}_i\right) = -\left(\alpha \rho\right)_i,$$  (10)

where the index $k$ takes values $\overline{1, \ N_{ch}}$, and the index $i$ takes value $\overline{1, \ N_{rc}}$.

Because of the neglected mass changes in the course of chemical reactions and mass transfer processes in the equilibrium case, it follows that the concentration of impurities non-radioactive fraction with high accuracy can be taken as constant, i. e. $\rho_k$=const. Then the system of equations (10) can be written as

$$\mathrm{div}\left(\vec{w}\right) = 0, \frac{\partial \rho_i}{\partial \tau} + \mathrm{div}\left(\vec{j}_i\right) = -\left(\alpha \rho\right)_i,$$  (11)

where - a vector velocity of the fluid.

Write out the flow $\vec{j}_i$ as the sum of two terms $\vec{j}_i = \vec{j}_{Di} + \rho_i \vec{v}$, where $\vec{j}_{Di}$ - the diffusion flux, $\rho_i \vec{w}$- the convective flow, and, taking into account the first equation (11), the second equation takes form

$$\frac{\partial \rho_i}{\partial \tau} + \mathrm{div}\left(\vec{j}_{Di}\right) + \vec{w} \cdot \nabla \rho_i = -\left(\alpha \rho\right)_i.$$  (12)

According to the Onsager linear theory, the flow for a multicomponent mixture can be written as follows

$$\vec{j}_{Di} = -\frac{L_{ij}}{T} \sum_k \left(\frac{\partial \mu_j}{\partial \rho_k}\right) \nabla \rho_k,$$  (13)

where $L_{ij}$ - the Onsager kinetic coefficients, $\mu_j$ - the chemical potential of $j$-th radionuclide.

The real radioactive solutions, arriving at the burial in a deep-seated formations, depending on the half-life have a total volumetric activity of about $10^{-6} \sim 10^1$ Ci/l. Let us estimate the mass of radionuclides in solution. Strontium-90 from the volumetric activity 1 Ci/l has a mass of about 7.57 $10^{-6}$ kg, and Ruthenium-106 is the same volumetric activity of the mass of the order of 0.3 $10^{-6}$ kg. These estimates of the mass of radionuclides provide a basis for considering solution under investigation to be a very dilute solution (with respect to radionuclide fractions). Therefore, the correlation between the diffusion fluxes of components j and k will be negligible.

Thus, the assumption of a very dilute solution leads to the following representation of (13)

$$\vec{j}_{Di} = -\frac{L_{ii}}{T}\left(\frac{\partial \mu_i}{\partial \rho_i}\right)\nabla \rho_i, \tag{14}$$

Introducing the notation $D_{ii} = \frac{L_{ii}}{T}\left(\frac{\partial \mu_i}{\partial \rho_i}\right)$, we obtain

$$\vec{j}_{Di} = -D_{ii}\nabla \rho_i. \tag{15}$$

Relation (15) is known as Fick's first law, where $D_{ii}$ is the diffusion coefficient of $i$-th - radionuclide.

In many cases, the diffusion coefficient $D_{ii}$ can be considered to be constant, then using (15) in equation (12), we obtain a system of equations for evolution of radionuclides in a porous layer

$$\frac{\partial \rho_i}{\partial \tau} - D_{ii}\Delta \rho_i + \vec{w}\cdot\nabla \rho_i = -\left(\alpha \rho\right)_i. \tag{16}$$

Equation (16) is written for the porous layer, but it does not take into account the presence of porosity and sorption of radionuclides in the skeleton of the formation. To account for these effects, we introduce an auxiliary space-time function $m = m(t, x, y, z)$, such that $\int m(t, x, y, z)dV = V_{por}$, where $V_{por}$ - the volume of pore space. Obviously, the $V_s = V - V_{por}$ - the volume occupied by the formation. Integrating each term of the of equation (16) by volume

$$\int\frac{\partial \rho_i}{\partial \tau}dV - D_{ii}\int \Delta \rho_i dV + \vec{w}\cdot\int\nabla \rho_i dV = -\int\left(\alpha \rho\right)_i dV. \tag{17}$$

Under the integral expression $\rho_i dV$ can be represented as the sum of two terms as $\rho_i dV = \rho_{is}dV_s + \rho_{ito}dV_{por}$, because the other terms, taking into account the mass transfer between the solution and the formation, give a zero contribution due to the steady equilibrium. Therefore

$$\int \frac{\partial \rho_{is}}{\partial \tau} dV_s + \int \frac{\partial \rho_{iw}}{\partial \tau} dV_{por} - D_{is}^i \int \Delta \rho_{is} dV_s - D_{iw}^i \int \Delta \rho_{iw} dV_{por} + \vec{w} \cdot \int \nabla \rho_{iw} dV_{por} =$$
$$= -\int (\alpha \rho_s)_i dV_s - \int (\alpha \rho_w)_i dV_{por}. \tag{18}$$

Using the definition of an auxiliary function $m$, it is easy to obtain the following obvious relations

$$dV_{por} = m dV, dV_s = (1 - m) dV. \tag{19}$$

Substituting (19) in equation (18), we obtain

$$\int (1-m) \frac{\partial \rho_{is}}{\partial \tau} dV + \int m \frac{\partial \rho_{iw}}{\partial \tau} dV - D_{is}^i \int (1-m) \Delta \rho_{is} dV - D_{iw}^i \int m \Delta \rho_{iw} dV +$$
$$+ \vec{w} \cdot \int m \nabla \rho_i dV = -\int (1-m)(\alpha \rho_s)_i dV - \int m(\alpha \rho_w)_i dV. \tag{20}$$

Again, because of the arbitrary choice of the reaction volume $dV$ and continuity of integrand functions, we obtain

$$(1-m) \frac{\partial \rho_{is}}{\partial \tau} + m \frac{\partial \rho_{iw}}{\partial \tau} - D_{is}^i (1-m) \Delta \rho_{is} - D_{iw}^i m \Delta \rho_{iw} + \vec{w} \cdot m \nabla \rho_{iw} = -(1-m)(\alpha \rho_s)_i - m(\alpha \rho_w)_i. \tag{21}$$

We assume that the dependence of the impurity concentration in the skeleton of its concentration in the fluid is linear (Henry's isotherm) and does not depend on the volume activity, that is a good approximation for relatively small concentrations of fraction of radionuclide

$$\rho_{is} = K_{iГ}^i \rho_{iw}. \tag{22}$$

Then the mass transfer equations take the form:

$$\left[ (1-m) K_{iГ}^i + m \right] \frac{\partial \rho_{iw}}{\partial \tau} - \left[ D_{is}^i (1-m) K_{iГ}^i + m D_{iw}^i \right] \Delta \rho_{iw} + \vec{w} \cdot m \nabla \rho_{iw} =$$
$$- \left[ (1-m) K_{iГ}^i + m \right] (\alpha \rho_w)_i, \tag{23}$$

where the function $m$- void factor, depending on lithological and mineral composition of the layer, $K_{iГ}^i$ - the Henry's coefficient of $i$-th - radionuclide.

The final form of the equations of evolution of radionuclides in solution (liquid phase) in a porous layer, taking into account the porosity and adsorption on the skeleton, one can divide both sides of equation (23) by a factor $\left((1-m)K_{il}^{i}+m\right)$

$$\frac{\partial \rho_{iw}}{\partial \tau} - D_{ii}\Delta\rho_{iw} + \vec{v}_{ii}'\cdot\nabla\rho_{iw} = -\left(\alpha\,\rho_{w}\right)_{i}, \tag{24}$$

here $D_{ii}=\left(D_{is}^{i}(1-m)K_{il}^{i}+m\,D_{iw}^{i}\right)/\left((1-m)K_{il}^{i}+m\right)$ - the effective diffusion coefficient in the layer, $\vec{v}_{ii}'=m\vec{w}/\left((1-m)K_{il}^{i}+m\right)$ - modified velocity of propagation of $i$-th - radionuclide in a porous layer, (the rate of convective transport of radioactive contaminants).

Note that equation (24) is derived for the case when a radionuclide decaying, forms a non-radioactive nuclide. Possible decay scheme

$A \rightarrow B \rightarrow C\ (stable),$

i. e. when the decay product $B$ will also be radioactive. The equation takes into account the formation of a child radionuclide

$$\frac{\partial \rho_{w}^{c}}{\partial \tau} - D\Delta\rho_{w}^{c} + \vec{v}\cdot\nabla\rho_{w}^{c} = \alpha\,\rho_{w} - \alpha^{c}\rho_{w}^{c}, \tag{25}$$

where $\rho_{w}^{c}$ - density of the child radionuclide, $\rho_{w}$ - density of the parent radionuclide. Investigations of these cases [36], [37] in the work are not included.

The rate of filtration of snap motion of the liquid phases is determined by Darcy's law

$$\vec{v} = -\frac{k}{\mu}\mathrm{grad}\,P.$$

In most common filtration processes, the deformation of the porous skeleton, compressibility, and associated with this changes in the temperature of liquids are small. The main effects that determine the motion of the system are the non-equlibrium joint motion of several liquid phases, molecular and convective diffusion of solute in the phases of the components, the absorption of the solid phase or sorption of the components, mass transfer between phases.

Thus, the system of equations describing the mass transfer during injection of liquid radioactive wastes in deep porous horizon is as follows:

$$\begin{aligned}
&\mathrm{div}\left(\vec{w}\right)=0,\\
&\frac{\partial \rho_{iw}}{\partial \tau} - D_{i}\Delta\rho_{iw} + \vec{v}_{i}'\cdot\nabla\rho_{iw} = -\alpha_{i}\,\rho_{iw},\\
&\frac{\partial \rho_{w}^{\tilde{n}}}{\partial \tau} - D\Delta\rho_{w}^{\tilde{n}} + \vec{v}\cdot\nabla\rho_{w}^{\tilde{n}} = \alpha\,\rho_{w} - \alpha^{\tilde{n}}\rho_{w}^{\tilde{n}}.
\end{aligned} \tag{26}$$

For a complete statement of the problem requires knowledge of the radiochemical composition of the solution, flow rate of the injection, diffusion parameters and the geometry of the simulated porous layer. Note that if the injection rate is known, it is easy to determine the rate of filtration. Then integrating Darcy's equation, we can describe the pressure field in the formation.

The problem under consideration has cylindrical symmetry about the axis of the well, through which the liquid wastes are ejected; it is convenient to represent the system of equations (26) in a cylindrical coordinate system.

Writing the first equation (26) in a cylindrical coordinate system and, given that the liquid is distributed in the porous layer only in the radial direction, we obtain the equation for the velocity field:

$$\frac{\partial(rw_r)}{\partial r_d} = 0,$$

solving this equation and applying the obvious boundary condition $w_r \mid_{r=r_0} = w_0$, where $w_0$ - velocity of the fluid from the cased hole in the porous layer, we have

$$w_r = w_0 r_0 / r_d. \tag{27}$$

Then the remaining equations of (26) using (27), and the anisotropy of diffusion coefficients and thermal conductivity in the directions $r$ and $z$ in a cylindrical coordinate system can be written as

$$\frac{\partial \rho_{iw}}{\partial \tau} + \frac{v'_0 r_0}{r_d} \frac{\partial \rho_{iw}}{\partial r_d} - D_{ri} \frac{1}{r_d} \frac{\partial}{\partial r_d} \left( r_d \frac{\partial \rho_{iw}}{\partial r_d} \right) - D_{zi} \frac{\partial^2 \rho_{iw}}{\partial z_d^2} = -\alpha_i \rho_{iw},$$

$$\frac{\partial \rho_w^{\tilde{n}}}{\partial \tau} + \frac{v'_0 r_0}{r_d} \frac{\partial \rho_w^{\tilde{n}}}{\partial r_F} - D_r \frac{1}{r_d} \frac{\partial}{\partial r_d} \left( r_d \frac{\partial \rho_w^c}{\partial r_d} \right) - D_z \frac{\partial^2 \rho_w^c}{\partial z_d^2} = \alpha \rho_w - \alpha \rho_w^c,$$
$$\tag{28}$$

where $v_0 = mw_0$ - the rate of fluid filtration, $D_{ri}$, $D_{zi}$ - an effective diffusion coefficient in the direction $r_d$ and $z_d$, respectively, and the multiplier $\gamma_i = [(1-m)K_{i\Gamma} + m]$.

It is assumed that the real porous layer is represented by a multiphase system, where each phase consists of a sufficiently large number of randomly distributed small particles. Particle size, small in comparison with the basic physical quantities are assumed to be so large that within each particle condition of "local equilibrium" [38] and all the conservation laws are satisfied [9]. All contact surfaces of particles of different nature are surfaces of discontinuity of some physical fields. However, the above assumptions allow us in physically small volumes to define the space of continuous functions, carrying out the description of the fields of each phase. This determination is carried out by a predetermined method of averaging, from which, in general, depend on the results obtained in [9]. As with most occurring filtration processes, the deformation of the porous skeleton, compressibility and associated changes in temperature fluids rely small.

Given that the determining factor in the process of mass transfer is the concentration of the parent nuclide, confine ourselves to the problem for a single pollutant, which is radioactive and chemically active. The first equation (28) is represented as

$$\frac{\partial \rho_w}{\partial \tau} + \frac{\vec{v} \nabla \rho_w}{(1-m)K_\Gamma + m} - D\Delta\rho_w = -\alpha\,\rho_w.$$
(29)

Here we have introduced the notation

$$D = \frac{D_s(1-m)K_r + D_w m}{(1-m)K_r + m}$$
(30)

$D$ is the effective diffusion coefficient in the layer. From (29) that in the equation describing the migration of contaminants, it is necessary to take into account the convective transport of pollutants, "complicated by" the presence of porosity in the skeleton and mass transfer processes occurring between the pollutant and the skeleton. Equation (29) to determine the rate of convective transport of pollutants in porous layer $\vec{v}'$, by analogy with the rate of convective heat transfer and flow rate $\vec{v}$

$$\vec{v}' = \frac{\vec{v}}{(1-m)K_\Gamma + m}.$$
(31)

The rate of convective transport of the impurity $\vec{v}'$ determines the position of the front of pollution $R_d$, just as the filtration rate $\vec{v}$ determines the position of the front of injected fluid $R_w$. The position of the injected fluid front is determined from the mass balance of the injected fluid and, for the case of injection at a constant speed $v_0$ into the layer through a cased hole of radius $r_0$, the corresponding expression is given by

$$R_w = \sqrt{\frac{2 v_0 r_0 \tau}{m} + r_0^2} = \sqrt{2 w_0 r_0 \tau + r_0^2} = \sqrt{\frac{Q\tau}{\pi m H} + r_0^2}.$$
(32)

## 3. The mathematical formulation of the problem of mass transfer

Fig. 1 shows the geometry of the problem in a cylindrical coordinate system whose axis coincides with the axis of the borehole. The environment is presented by three areas with flat boundaries. Injection of impurities into the area is out of the hole radius, covering and underlying layers are impermeable, middle area is a the porous region, all layers are considered homogeneous and anisotropic on the diffusion properties. Observation is carried out at a distance from the axis of the borehole

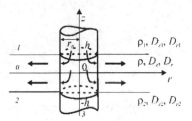

**Figure 1.** The geometry of the problem: 0, 1, 2 - porous, covering and underlying layers, respectively, 3 – borehole

Through a hole of small (compared to the distance to the observation point) radius $r_0$ in an infinite horizontal layer of thickness $-h < z_d < h$ water with a radioactive contaminant are injected. In arriving liquid at $r \leq r_0$ the concentration of impurities kept constant and equal to $\rho_0$. The concentration of pollutants in the layer changes due to convective transport along the direction $r$, the diffusion along $r$, $z$ and the concentration of sources. As such sources of radioactive decay of pollutant are considered. Field of densities during the filtration of radioactive solutions was investigated in [36, 39-50].

The mathematical formulation of the problem of mass transfer for all areas involves the diffusion equation with taking into account the radioactive decay in the covering

$$\frac{\partial \rho_{1d}}{\partial \tau} - D_{z1}\frac{\partial^2 \rho_{1d}}{\partial z_d^2} - D_{r1}\frac{1}{r_d}\frac{\partial}{\partial r_d}\left(r_d \frac{\partial \rho_{1d}}{\partial r_d}\right) = -\alpha\rho_{1d}, \ \tau > 0, \ r_d > 0, \ z_d > h \tag{33}$$

and the underlying

$$\frac{\partial \rho_{2d}}{\partial \tau} - D_{z2}\frac{\partial^2 \rho_{2d}}{\partial z_d^2} - D_{r2}\frac{1}{r_d}\frac{\partial}{\partial r_d}\left(r_d \frac{\partial \rho_{2d}}{\partial r_d}\right) = -\alpha\rho_{2d}, \ \tau > 0, \ r_d > 0, \ z_d < -h \tag{34}$$

layers, as well as the equation of convective diffusion, taking into account the radioactive decay in the porous layer

$$\frac{\partial \rho_d}{\partial \tau} - D_z\frac{\partial^2 \rho_d}{\partial z_d^2} - D_r\frac{1}{r_d}\frac{\partial}{\partial r_d}\left(r_d \frac{\partial \rho_d}{\partial r_d}\right) + \frac{v_0' r_0}{r_d}\frac{\partial \rho_d}{\partial r_d} = -\alpha\rho_d, \ \tau > 0, \ r_d > 0, \ |z_d| < h. \tag{35}$$

The conditions of conjugation represent the equality of densities and fluxes of dissolved substances at the interface of the layers

$$\rho_d\big|_{z_d=h} = \rho_{d1}\big|_{z_d=h} \, , \rho_d\big|_{z_d=-h} = \rho_{d2}\big|_{z_d=-h} \, , \tag{36}$$

$$D_z\frac{\partial \rho_d}{\partial z_d}\bigg|_{z_d=h} = D_{z1}\frac{\partial \rho_{1d}}{\partial z_d}\bigg|_{z_d=h} \, , D_z\frac{\partial \rho_d}{\partial z_d}\bigg|_{z_d=-h} = D_{z2}\frac{\partial \rho_{2d}}{\partial z_d}\bigg|_{z_d=-h} \, . \tag{37}$$

The density of pollutant at the entrance of porous layer assumed to be constant

$$\rho_d\big|_{r_d=0} = \big[m + K(1-m)\big]\rho_0 \, . \tag{38}$$

Assuming that at the initial time the density of the of pollutant is equal to zero

$$\rho_d\big|_{\tau=0} = \rho_{1d}\big|_{\tau=0} = \rho_{2d}\big|_{\tau=0} = 0. \tag{39}$$

In addition, at infinity the conditions of regularity

$$\rho_d\big|_{r_d \to +\infty} = 0, \, \rho_{1d}\big|_{r_d+z_d \to +\infty} = 0, \, \rho_{2d}\big|_{r_d+|z_d| \to +\infty} = 0. \tag{40}$$

Let us turn then to the dimensionless quantities

$$\rho = \frac{\rho_d}{\rho_0}, \quad r = \frac{r_d}{h}, \quad z = \frac{z_d}{h}, \quad t = \frac{\lambda_{z1}\tau}{c_1\rho_{n1}h^2}, \quad \text{At} = \frac{c_1\rho_{n1}}{\lambda_{z1}}\alpha h^2, \quad {}_1^2 D = \frac{D_{z2}}{D_{z1}},$$

$$ {}_1^0 D = \frac{D_z}{D_{z1}}, \quad {}_2^0 D = \frac{D_z}{D_{z2}}, \quad a_{z1} = \frac{\lambda_{z1}}{c_1\rho_{n1}}, \quad \gamma = \frac{1}{m+K(1-m)} \, . $$

We also introduce the analogue of the Péclet number

$$\text{Pd} = v'_0 r_0 / D_{z1},$$

where $v'_0$ is the rate of convective transport of the pollutant at a distance $r_0$ from the axis of the borehole. With this notation equations (33) - (40) take the form

$$\frac{\partial \rho_1}{\partial t} - \frac{\partial^2 \rho_1}{\partial z^2} - \frac{D_{r1}}{D_{z1}}\frac{1}{r}\frac{\partial}{\partial r}\left(r\frac{\partial \rho_1}{\partial r}\right) = -\text{At}\,\rho_1, \quad t > 0, \quad r > 0, \quad z > 1, \tag{41}$$

$$\frac{\partial \rho_2}{\partial t} - {}_i^2 D \frac{\partial^2 \rho_2}{\partial z^2} - \frac{D_{r2}}{D_{z1}} \frac{1}{r} \frac{\partial}{\partial r}\left(r \frac{\partial \rho_2}{\partial r}\right) = -At\, \rho_2, \quad t>0, \quad r>0, \quad z<-1, \tag{42}$$

$$\frac{\partial \rho}{\partial t} - {}_i^0 D \frac{\partial^2 \rho}{\partial z^2} - \frac{D_r}{D_{z1}} \frac{1}{r} \frac{\partial}{\partial r}\left(r \frac{\partial \rho}{\partial r}\right) + \frac{Pd}{r} \frac{\partial \rho}{\partial r} = -At\, \rho, \quad t>0, r>0, |z|<1. \tag{43}$$

Let us estimate the ratio of the third and fourth terms in equation (43)

$$\frac{\dfrac{D_r}{D_{z1}} \dfrac{1}{r} \dfrac{\partial}{\partial r}\left(r \dfrac{\partial \rho}{\partial r}\right)}{\dfrac{Pd}{r} \dfrac{\partial \rho}{\partial r}} \approx \frac{D_r \dfrac{\rho_0}{R^2}}{D_{z1} Pd \dfrac{\rho_0}{R^2}} = \frac{D_r}{D_{z1} Pd} \approx \frac{1}{Pd} <<1,$$

Boundary, initial conditions and conjugation conditions are not changed

$$\rho\big|_{t=0} = \rho_1\big|_{t=0} = \rho_2\big|_{t=0} = 0 \ , \tag{44}$$

$$\rho\big|_{r=0} = 1/\gamma, \quad \rho\big|_{z=1} = \rho_1\big|_{z=1}, \quad \rho\big|_{z=-1} = \rho_2\big|_{z=-1}, \tag{45}$$

$$\frac{\partial \rho}{\partial z}\bigg|_{z=1} = {}_0^1 D \frac{\partial \rho_1}{\partial z}\bigg|_{z=1}, \quad \frac{\partial \rho}{\partial z}\bigg|_{z=-1} = {}_0^2 D \frac{\partial \rho_2}{\partial z}\bigg|_{z=-1}, \tag{46}$$

$$\rho\big|_{r\to\infty} = 0, \quad \rho_1\big|_{r+z\to\infty} = 0, \quad \rho_2\big|_{r+|z|\to\infty} = 0. \tag{47}$$

The system of equations (44) - (50) defines a mathematical formulation of the problem of mass transfer.

## 4. Expansion of the solution to the problem of mass transfer on the asymptotic parameter

Let us consider the more general problem, which is obtained by introducing into the equations and boundary conditions of arbitrary asymptotic parameter $\varepsilon$ of formal substitution in the diffusion coefficient $D_z$ for $D_z/\varepsilon$. In accordance with the designations this performed by replacing ${}_0^1 D$ for $\varepsilon\, {}_0^1 D$ and ${}_0^2 D = {}_1^2 D {}_0^1 D$ for $\varepsilon\, {}_0^2 D$. Note that the original problem can be obtained from the solution of a parameterized the problem when $\varepsilon = 1$. The problem (44) - (50) is thus a particular case of the more general parameterized problem, containing a parameter of the asymptotic expansion both in the equation for the layer and in the conditions of conjugation

$$\frac{\partial \rho_1}{\partial t} - \frac{\partial^2 \rho_1}{\partial z^2} = -At \; \rho_1, \quad t > 0, \quad r > 0, \quad z > 1, \tag{51}$$

$$\varepsilon \frac{\partial \rho}{\partial t} - {}_1^0 D \frac{\partial^2 \rho}{\partial z^2} + \varepsilon \frac{Pd}{r} \frac{\partial \rho}{\partial r} = -\varepsilon At \rho, \quad t > 0, \quad r > 0, \quad |z| < 1, \tag{52}$$

$$\frac{\partial \rho_2}{\partial t} - {}_1^2 D \frac{\partial^2 \rho_2}{\partial z^2} = -At \rho_2, \quad t > 0, \quad r > 0, \quad z < -1 \tag{53}$$

with boundary conditions

$$\left. \frac{\partial \rho}{\partial z} \right|_{z=1} = {}_0^1 D \varepsilon \left. \frac{\partial \rho_1}{\partial z} \right|_{z=1}, \quad \left. \frac{\partial \rho}{\partial z} \right|_{z=-1} = {}_0^2 D \varepsilon \left. \frac{\partial \rho_2}{\partial z} \right|_{z=-1}, \tag{54}$$

$$\rho \big|_{r=0} = 1/\gamma, \rho \big|_{z=1} = \rho_1 \big|_{z=1}, \quad \rho \big|_{z=-1} = \rho_2 \big|_{z=-1}, \tag{55}$$

$$\rho \big|_{t=0} = \rho_1 \big|_{t=0} = \rho_2 \big|_{t=0} = 0, \tag{56}$$

$$\rho \big|_{r \to +\infty} = 0, \rho_1 \big|_{r+z \to +\infty} = 0, \rho_2 \big|_{r+|z| \to +\infty} = 0. \tag{57}$$

To find the solution to (51) - (57), one can represent the density function $\rho$ of each region by the asymptotic formula of the parameter $\varepsilon$

$$\rho = \rho^{(0)} + \varepsilon \rho^{(1)} + \ldots + \varepsilon^n \rho^{(n)} + \theta^{(n)}, \rho_i = \rho_i^{(0)} + \varepsilon \rho_i^{(1)} + \ldots + \varepsilon^n \rho_i^{(n)} + \theta_i^{(n)}, i = 1,2. \tag{58}$$

Substituting expression (58) in (51) - (57) and grouping terms in powers of the expansion parameter $\varepsilon$, one can easily obtain

$$\left( \frac{\partial \rho_1^{(0)}}{\partial t} - \frac{\partial^2 \rho_1^{(0)}}{\partial z^2} + At \rho_1^{(0)} \right) + \varepsilon \left( \frac{\partial \rho_1^{(1)}}{\partial t} - \frac{\partial^2 \rho_1^{(1)}}{\partial z^2} + At \rho_1^{(1)} \right) + \ldots = 0, \; t > 0, \quad r > 0, \quad z > 1, \tag{59}$$

$$\left(\frac{\partial \rho_2^{(0)}}{\partial t} - {}_1^2D\frac{\partial^2 \rho_2^{(0)}}{\partial z^2} + At\rho_2^{(0)}\right) + \varepsilon\left(\frac{\partial \rho_2^{(1)}}{\partial t} - {}_1^2D\frac{\partial^2 \rho_2^{(1)}}{\partial z^2} + At\rho_2^{(1)}\right) + ... = 0, \, t > 0, \, r > 0, \, z < -1, \quad (60)$$

$$-{}_1^0D\frac{\partial^2 \rho^{(0)}}{\partial z^2} + \varepsilon\left(\frac{\partial \rho^{(0)}}{\partial t} - {}_1^0D\frac{\partial^2 \rho^{(1)}}{\partial z^2} + \frac{Pd}{r}\frac{\partial \rho^{(0)}}{\partial r} + At\rho^{(0)}\right) +$$

$$+\varepsilon^2\left(\frac{\partial \rho^{(1)}}{\partial t} - {}_1^0D\frac{\partial^2 \rho^{(2)}}{\partial z^2} + \frac{Pd}{r}\frac{\partial \rho^{(1)}}{\partial r} + At\rho^{(1)}\right) + ... = 0, \quad t > 0, \, r > 0, \, |z| < 1, \quad (61)$$

$$\frac{\partial \rho^{(0)}}{\partial z}\bigg|_{z=1} + \varepsilon\left(\frac{\partial \rho^{(1)}}{\partial z}\bigg|_{z=1} - {}_0^1D\frac{\partial \rho_1^{(0)}}{\partial z}\bigg|_{z=1}\right) + ... = 0, \frac{\partial \rho^{(0)}}{\partial z}\bigg|_{z=-1} + \varepsilon\left(\frac{\partial \rho^{(1)}}{\partial z}\bigg|_{z=-1} - {}_0^2D\frac{\partial \rho_2^{(0)}}{\partial z}\bigg|_{z=-1}\right) + ... = 0, \quad (62)$$

$$\left(\rho^{(0)} + \varepsilon\rho^{(1)} + ...\right)\bigg|_{z=1} = \left(\rho_1^{(0)} + \varepsilon\rho_1^{(1)} + ...\right)\bigg|_{z=1}, \left(\rho^{(0)} + \varepsilon\rho^{(1)} + ...\right)\bigg|_{z=-1} = \left(\rho_2^{(0)} + \varepsilon\rho_2^{(1)} + ...\right)\bigg|_{z=-1}, \quad (63)$$

$$\left(\rho^{(0)} + \varepsilon\rho^{(1)} + ...\right)\bigg|_{t=0} = \left(\rho_1^{(0)} + \varepsilon\rho_1^{(1)} + ...\right)\bigg|_{t=0} = \left(\rho_2^{(0)} + \varepsilon\rho_2^{(1)} + ...\right)\bigg|_{t=0} = 0, \quad (64)$$

$$\left(\rho^{(0)} + \varepsilon\rho^{(1)} + ...\right)\bigg|_{r=0} = 1/\gamma, \quad (65)$$

$$\left(\rho^{(0)} + \varepsilon\rho^{(1)} + ...\right)\bigg|_{r\to+\infty} = 0, \left(\rho_1^{(0)} + \varepsilon\rho_1^{(1)} + ...\right)\bigg|_{r+z\to+\infty} = 0, \left(\rho_2^{(0)} + \varepsilon\rho_2^{(1)} + ...\right)\bigg|_{r+|z|\to+\infty} = 0. \quad (66)$$

Analysis of the formulation of the problem shows that the factors of powers of $\varepsilon$ in (61) contain the neighboring coefficients of the expansion, and in this sense, are linked. To solve the corresponding equations implemented decoupling procedure.

### 4.1. The mathematical formulation of the problem of mass transfer in a zero approximation

If we formally consider $\varepsilon$ in equation (61) to be infinitely small, we obtain ${}_1^0D\partial^2\rho^{(0)}/\partial z^2 = 0$.

The result of integration $\partial\rho^{(0)}/\partial z = A(r, t)$ with the boundary conditions (62) allows us to establish that $A(r, t) = 0$. Thus, in the zero approximation, the density of the pollutant $\rho^{(0)} = \rho^{(0)}(r, t)$ is a function only of $r$ and $t$. Consequently, in the zero approximation the density

of the pollutant in each cylindrical cross section with the axis z is the same in height of the
carrier layer. Next, equating to zero coefficients near $\varepsilon$ in equation (61), we obtain

$$\frac{\partial \rho^{(0)}}{\partial t} - {}_1^0 D \frac{\partial^2 \rho^{(1)}}{\partial z^2} + \frac{Pd}{r} \frac{\partial \rho^{(0)}}{\partial r} + At \, \rho^{(0)} = 0. \tag{67}$$

Since $\rho^{(0)}(r, t)$ does not depend on z, the auxiliary function $E(r, t)$, composed of the terms of
the equation (67) containing $\varrho^{(0)}$

$$E(r,t) = \frac{\partial \rho^{(0)}}{\partial t} + \frac{Pd}{r} \frac{\partial \rho^{(0)}}{\partial r} + At \, \rho^{(0)}, \tag{68}$$

is also independent of z. Then (1) can be written as

$${}_1^0 D \frac{\partial^2 \rho^{(1)}}{\partial z^2} = E(r,t). \tag{69}$$

Integrating successively, one can find the expression for the first derivative of the first
coefficient $\rho^{(1)}$ of the variable z

$$\frac{\partial \rho^{(1)}}{\partial z} = {}_0^I D \left[ z E(r,t) + F(r,t) \right], \tag{70}$$

and the first coefficient of expansion in the form of a quadratic trinomial

$$\rho^{(1)} = {}_0^I D \left( \frac{z^2}{2} E(r,t) + z F(r,t) + Q(r,t) \right), \tag{71}$$

with the functional coefficients to be determined. From the boundary conditions (62) with the
cofactor $\varepsilon$ we have

$${}_i^2 D \frac{\partial \rho_1^{(0)}}{\partial z} \bigg|_{z=1} = E(r,t) + F(r,t), \, {}_i^2 D \frac{\partial \rho_2^{(0)}}{\partial z} \bigg|_{z=-1} = -E(r,t) + F(r,t). \tag{72}$$

Hence, one can obtain an expression for the functional coefficients and through the traces of derivatives of the outer regions

$$E(r,t) = \frac{1}{2}\left( \left.\frac{\partial \rho_1^{(0)}}{\partial z}\right|_{z=1} - {}_i^2 D \left.\frac{\partial \rho_2^{(0)}}{\partial z}\right|_{z=-1} \right), \tag{73}$$

$$F(r,t) = \frac{1}{2}\left( \left.\frac{\partial \rho_1^{(0)}}{\partial z}\right|_{z=1} + {}_i^2 D \left.\frac{\partial \rho_2^{(0)}}{\partial z}\right|_{z=-1} \right). \tag{74}$$

Substituting (73) in (74), one can obtain the desired equation for the zero approximation of the density of impurities in the layer

$$\frac{\partial \rho^{(0)}}{\partial t} + \frac{Pd}{r}\frac{\partial \rho^{(0)}}{\partial t} + At\rho^{(0)} = \frac{1}{2}\left( \left.\frac{\partial \rho_1^{(0)}}{\partial z}\right|_{z=1} - {}_i^2 D \left.\frac{\partial \rho_2^{(0)}}{\partial z}\right|_{z=-1} \right). \tag{75}$$

The final statement of the problem in the zero approximation also includes the equations in the covering and underlying formations

$$\frac{\partial \rho_1^{(0)}}{\partial t} - \frac{\partial^2 \rho_1^{(0)}}{\partial z^2} = -At\rho_1^{(0)}, \quad t>0, \ r>0, \ z>1, \tag{76}$$

$$\frac{\partial \rho_2^{(0)}}{\partial t} - {}_i^2 D \frac{\partial^2 \rho_2^{(0)}}{\partial z^2} = -At\rho_2^{(0)}, \quad t>0, \ r>0, \ z<-1, \tag{77}$$

$$\frac{\partial \rho^{(0)}}{\partial t} + \frac{Pd}{r}\frac{\partial \rho^{(0)}}{\partial r} + At\rho^{(0)} = \frac{1}{2}\left( \left.\frac{\partial \rho_1^{(0)}}{\partial z}\right|_{z=1} - {}_i^2 D \left.\frac{\partial \rho_2^{(0)}}{\partial z}\right|_{z=-1} \right), t>0, \ r>0, \ |z|<1, \tag{78}$$

and the appropriate initial and boundary conditions

$$\left.\rho^{(0)}\right|_{t=0} = \left.\rho_1^{(0)}\right|_{t=0} = \left.\rho_2^{(0)}\right|_{t=0} = 0, \tag{79}$$

$$\rho^{(0)} = \rho_1^{(0)}\Big|_{z=1} = \rho_2^{(0)}\Big|_{z=-1}, \tag{80}$$

$$\rho^{(0)}\Big|_{r=0} = 1/\gamma, \tag{81}$$

$$\rho^{(0)}\Big|_{r\to+\infty} = 0, \rho_1^{(0)}\Big|_{r+z\to+\infty} = 0, \rho_2^{(0)}\Big|_{r+|z|\to+\infty} = 0. \tag{82}$$

Expressions (76) - (82) represent the boundary value problem for zero expansion coefficient $\rho^{(0)}$ or zero approximation. Note that in contrast to the original, which is the problem of conjugation for parabolic equations, it is mixed, since the equation contains traces of derivatives from the outer regions.

Finding the zero approximation of the density of the radioactive contaminant is important because just that approach arises in the zero approximation of the temperature problem.

### 4.2. The zero approximation of the problem of mass transport as the solution of the averaged problem

Let us average a parameterized problem (51) - (57) over z within the carrier layer according to

$$\langle\rho\rangle = \frac{1}{2}\int_{-1}^{1}\rho\,dz.$$

Successively averaging each term of equation (34)

$$\left\langle\frac{\partial\rho}{\partial t}\right\rangle = \frac{1}{2}\int_{-1}^{1}\frac{\partial\rho}{\partial t}\,dz = \frac{\partial\langle\rho\rangle}{\partial t}, \quad \mathrm{Pd}\frac{1}{r}\left\langle\frac{\partial\rho}{\partial r}\right\rangle = \mathrm{Pd}\frac{1}{r}\frac{\partial\langle\rho\rangle}{\partial r}$$

$$\frac{1}{\varepsilon}\cdot {}_1^0 D\left\langle\frac{\partial^2\rho}{\partial z^2}\right\rangle = \frac{{}_1^0 D}{2\varepsilon}\int_{-1}^{1}\frac{\partial^2\rho}{\partial z^2}dz = \frac{{}_1^0 D}{2\varepsilon}\left(\frac{\partial\rho}{\partial z}\Big|_{z=1} - \frac{\partial\rho}{\partial z}\Big|_{z=-1}\right) = \frac{1}{2}\left(\frac{\partial\rho_1}{\partial z}\Big|_{z=1} - {}_1^2 D\frac{\partial\rho_2}{\partial z}\Big|_{z=-1}\right),$$

one can obtain the following formulation of the averaged parameterized problem:

$$\frac{\partial\rho_1}{\partial t} - \frac{\partial^2\rho_1}{\partial z^2} + \mathrm{At}\,\rho_1 = 0, \quad t>0, \quad r>0, \quad z>1, \tag{83}$$

$$\frac{\partial\langle\rho\rangle}{\partial t} - \frac{1}{2}\left(\frac{\partial\rho_1}{\partial z}\Big|_{z=1} - {}_1^2 D\frac{\partial\rho_2}{\partial z}\Big|_{z=-1}\right) + \mathrm{Pd}\frac{1}{r}\frac{\partial\langle\rho\rangle}{\partial r} = -\mathrm{At}\langle\rho\rangle, \quad t>0, \quad r>0, \quad |z|<1, \tag{84}$$

$$\frac{\partial \rho_2}{\partial t} - {}_1^2 D \frac{\partial^2 \rho_2}{\partial z^2} + At\,\rho_2 = 0, \quad t>0, \quad r>0, \quad z<-1 \tag{85}$$

$$\langle \rho \rangle = \rho_1\big|_{z=1} = \rho_2\big|_{z=-1}, \tag{86}$$

$$\langle \rho \rangle\big|_{r=0} = 1/\gamma, \tag{87}$$

$$\langle \rho \rangle\big|_{t=0} = \rho_1\big|_{t=0} = \rho_2\big|_{t=0} = 0, \tag{88}$$

$$\langle \rho \rangle\big|_{r\to+\infty} = 0, \rho_1\big|_{r+z\to+\infty} = 0, \rho_2\big|_{r+|z|\to+\infty} = 0. \tag{89}$$

The resulting problem coincides with problem (83) - (89) for the zero approximation of the density of pollutant. The uniqueness of solutions implies that $\langle \rho \rangle = \rho^{(0)}$, i. e. zero approximation describes the known way averaged solutions to the original problem. If average the original nonparameterized problem (44) - (50), it also coincides with the problem for zero approximation of the field densities of the pollutant.

### 4.3. The mathematical formulation of the problem of mass transfer in the first approximation

Equations (41) - (43) for the coefficients for $\varepsilon$ take the form

$$\frac{\partial \rho_1^{(1)}}{\partial t} - \frac{\partial^2 \rho_1^{(1)}}{\partial z^2} + At\,\rho_1^{(1)} = 0, \quad t>0, \ r>0, \ z>1, \tag{90}$$

$$\frac{\partial \rho_2^{(1)}}{\partial t} - {}_1^2 D \frac{\partial^2 \rho_2^{(1)}}{\partial z^2} + At\,\rho_2^{(1)} = 0, \quad t>0, \ r>0, \ z<-1, \tag{91}$$

$$\frac{\partial \rho^{(1)}}{\partial t} + \frac{Pd}{r}\frac{\partial \rho^{(1)}}{\partial r} - {}_1^0 D \frac{\partial^2 \rho^{(2)}}{\partial z^2} + At\,\rho^{(1)} = 0, \quad t>0, \ r>0, \ |z|<1, \tag{92}$$

appropriate boundary and initial conditions are represented as

$$\frac{\partial \rho^{(1)}}{\partial z}\bigg|_{z=1} - \frac{1}{0}D\frac{\partial \rho_1^{(0)}}{\partial z}\bigg|_{z=1} = 0, \frac{\partial \rho^{(1)}}{\partial z}\bigg|_{z=-1} - \frac{2}{0}D\frac{\partial \rho_2^{(0)}}{\partial z}\bigg|_{z=-1} = 0, \tag{93}$$

$$\rho^{(1)}\bigg|_{z=1} = \rho_1^{(1)}\bigg|_{z=1}, \rho^{(1)}\bigg|_{z=-1} = \rho_2^{(1)}\bigg|_{z=-1}, \tag{94}$$

$$\rho^{(1)}\bigg|_{t=0} = \rho_1^{(1)}\bigg|_{t=0} = \rho_2^{(1)}\bigg|_{t=0} = 0, \tag{95}$$

$$\rho^{(1)}\bigg|_{r\to+\infty} = 0, \rho_1^{(1)}\bigg|_{r+z\to+\infty} = 0, \rho_2^{(1)}\bigg|_{r+|z|\to+\infty} = 0, \tag{96}$$

$$\rho^{(1)}\bigg|_{r=0} = 0. \tag{97}$$

The solution to the problem is sought in the form of quadratic polynomial for $z$ according to (71), where the auxiliary functions $E(r, t)$ and $F(r, t)$ are defined by (73) - (74), and the function $Q(r, t)$ is unknown. For its determination we can write (92) as

$$\frac{\partial^2 \rho^{(2)}}{\partial z^2} = \frac{1}{0}\frac{D}{\delta}\hat{L}\rho^{(1)}, \tag{98}$$

here $\hat{L}$ is the operator

$$\hat{L} = \frac{\partial}{\partial t} + \frac{Pd}{r}\frac{\partial}{\partial r} + At. \tag{99}$$

Using (71) and (98), and linearity of the operator $\hat{L}$, one can obtain

$$\frac{\partial^2 \rho^{(2)}}{\partial z^2} = \frac{1}{0}\frac{D^2}{\delta^2}\left(\frac{z^2}{2}\hat{L}E(r,t) + z\hat{L}F(r,t) + \hat{L}Q(r,t)\right). \tag{100}$$

Integrating the last expression over $z$

$$\frac{\partial \rho^{(2)}}{\partial z} = \frac{{}_0^1 D^2}{\delta^2} \left( \frac{z^3}{6} \hat{L} E(r,t) + \frac{z^2}{2} \hat{L} F(r,t) + z \hat{L} Q(r,t) + W(r,t) \right). \tag{101}$$

From the expression (101) and the boundary conditions (62) we have

$$\left. {}_I^0 D \frac{\partial \rho_1^{(1)}}{\partial z} \right|_{z=1} = \frac{1}{6} \hat{L} E(r,t) + \frac{1}{2} \hat{L} F(r,t) + \hat{L} Q(r,t) + W(r,t), \tag{102}$$

$$\left. {}_I^0 D {}_I^2 D \frac{\partial \rho_2^{(1)}}{\partial z} \right|_{z=-1} = -\frac{1}{6} \hat{L} E(r,t) + \frac{1}{2} \hat{L} F(r,t) - \hat{L} Q(r,t) + W(r,t). \tag{103}$$

From (90) and (91) one can get the equation for the definition of $Q$

$$\hat{L} Q(r,t) = \frac{{}_I^0 D}{2} \left( \left. \frac{\partial \rho_1^{(1)}}{\partial z} \right|_{z-1} - {}_I^2 D \left. \frac{\partial \rho_2^{(1)}}{\partial z} \right|_{z=-1} \right) - \frac{1}{6} \hat{L} E(r,t). \tag{104}$$

The equation for the determination of first coefficient of expansion is obtained by substituting (57), (58) and (55) in (98) with using (100)

$$\hat{L} \rho^{(1)} = \frac{{}_1^1 D}{4} \left( z^2 - \frac{1}{3} \right) \hat{L} \left( \left. \frac{\partial \rho_1^{(0)}}{\partial z} \right|_{z=1} - {}_I^2 D \left. \frac{\partial \rho_2^{(0)}}{\partial z} \right|_{z=-1} \right) +$$
$$\frac{{}_0^1 D}{2} z \hat{L} \left( \left. \frac{\partial \rho_1^{(0)}}{\partial z} \right|_{z=1} + {}_I^2 D \left. \frac{\partial \rho_2^{(0)}}{\partial z} \right|_{z=-1} \right) + \frac{1}{2} \left( \left. \frac{\partial \rho_1^{(1)}}{\partial z} \right|_{z=1} - {}_I^2 D \left. \frac{\partial \rho_2^{(1)}}{\partial z} \right|_{z=-1} \right). \tag{105}$$

The problem to determine the first coefficient of expansion also includes equations (90), (91) for the density field in the covering and underlying layers, respectively.

It is easy to see that the problem formulated by (90), (91), (105), (93) - (96) with the condition (97) has a trivial solution, so the condition (97) is redundant and should be weakened.

### 4.4. The problem for the remainder — Additional boundary condition

Substituting the expansion (58) with $n = 1$ in the parameterized problem (51) - (57), we obtain a problem for the remainder term

$$\frac{\partial \theta_1}{\partial t} - \frac{\partial^2 \theta_1}{\partial z^2} = -At\theta_1, \quad t > 0, \ r > 0, \ z > 1, \tag{106}$$

$$\varepsilon \frac{\partial \theta}{\partial t} - {}_1^0 D \frac{\partial^2 \theta}{\partial z^2} + \varepsilon \frac{Pd}{r} \frac{\partial \theta}{\partial r} + \varepsilon \, At\theta = -\varepsilon^2 \hat{L} \, \rho^{(1)}, \quad t > 0, \ r > 0, \ |z| < 1, \tag{107}$$

$$\frac{\partial \theta_2}{\partial t} - {}_i^2 D \frac{\partial^2 \theta_2}{\partial z^2} = -At\theta_2, \quad t > 0, \ r > 0, \ z < -1 \tag{108}$$

with the boundary conditions and conjugation conditions

$$\frac{\partial \theta}{\partial z}\bigg|_{z=1} = {}_0^1 D \varepsilon \frac{\partial \theta_1}{\partial z}\bigg|_{z=1} + {}_0^1 D \varepsilon^2 \frac{\partial \rho_1^{(1)}}{\partial z}\bigg|_{z=1}, \frac{\partial \theta}{\partial z}\bigg|_{z=-1} = {}_0^2 D \varepsilon \frac{\partial \theta_2}{\partial z}\bigg|_{z=-1} + {}_0^2 D \varepsilon^2 \frac{\partial \rho_2^{(1)}}{\partial z}\bigg|_{z=-1} \tag{109}$$

$$\theta\big|_{z=1} = \theta_1\big|_{z=1}, \quad \theta\big|_{z=-1} = \theta_2\big|_{z=-1}, \tag{110}$$

$$\theta\big|_{t=0} = \theta_1\big|_{t=0} = \theta_2\big|_{t=0} = 0, \tag{111}$$

$$\theta\big|_{r=0} = -\varepsilon \rho^{(1)}\big|_{r=0}, \tag{112}$$

$$\theta\big|_{r \to +\infty} = 0, \theta_1\big|_{r+z \to +\infty} = 0, \theta_2\big|_{r+|z| \to +\infty} = 0. \tag{113}$$

Restrict our investigation of the problem, averaged over the thickness of the layer. By averaging the second derivative over the vertical coordinate, use the conjugation conditions (109)

$$\frac{{}_1^0 D}{\varepsilon}\left\langle \frac{\partial^2 \theta}{\partial z^2} \right\rangle = \frac{{}_1^0 D}{2\varepsilon} \int_{-1}^{1} \frac{\partial^2 \theta}{\partial z^2} dz = \frac{{}_1^0 D \delta}{2\varepsilon}\left( \frac{\partial \theta}{\partial z}\bigg|_{z=1} - \frac{\partial \theta}{\partial z}\bigg|_{z=-1} \right) =$$

$$\frac{1}{2}\left( \frac{\partial \theta_1}{\partial z}\bigg|_{z=1} - {}_1^2 D \frac{\partial \theta_2}{\partial z}\bigg|_{z=-1} \right) + \frac{\varepsilon}{2}\left( \frac{\partial \rho_1^{(1)}}{\partial z}\bigg|_{z=1} - {}_1^2 D \frac{\partial \rho_2^{(1)}}{\partial z}\bigg|_{z=-1} \right).$$

The final formulation of the averaged problem for the remainder term represented as

$$\frac{\partial \theta_1}{\partial t} - \frac{\partial^2 \theta_1}{\partial z^2} = -At\theta_1, \quad t > 0, \ r > 0, \ z > 1, \tag{114}$$

$$\frac{\partial \langle \theta \rangle}{\partial t} + \frac{Pd}{r} \frac{\partial \langle \theta \rangle}{\partial r} + At\langle \theta \rangle - \frac{1}{2} \left( \frac{\partial \theta_1}{\partial z} \bigg|_{z=1} - {}_i^2 D \frac{\partial \theta_2}{\partial z} \bigg|_{z=-1} \right) =$$
$$= -\varepsilon \left( \hat{L} \langle \rho^{(1)} \rangle - \frac{1}{2} \left( \frac{\partial \rho_1^{(1)}}{\partial z} \bigg|_{z=1} - {}_i^2 D \frac{\partial \rho_2^{(1)}}{\partial z} \bigg|_{z=-1} \right) \right), \quad t > 0, \ r > 0, \ |z| < 1, \tag{115}$$

$$\frac{\partial \theta_2}{\partial t} - {}_i^2 D \frac{\partial^2 \theta_2}{\partial z^2} = -At\theta_2, \quad t > 0, \ r > 0, \ z < -1, \tag{116}$$

$$\langle \theta \rangle = \theta_1 \big|_{z=1} = \theta_2 \big|_{z=-1}, \tag{117}$$

$$\langle \theta \rangle \big|_{r=0} = -\varepsilon \langle \rho^{(1)} \rangle \big|_{r=0}, \tag{118}$$

$$\langle \theta \rangle \big|_{t=0} = \theta_1 \big|_{t=0} = \theta_2 \big|_{t=0} = 0, \tag{119}$$

$$\langle \theta \rangle \big|_{r \to +\infty} = 0, \theta_1 \big|_{r+z \to +\infty} = 0, \theta_2 \big|_{r+|z| \to +\infty} = 0 \tag{120}$$

It is easy to show that the averaged problem (110) - (116) for the remainder term has a trivial solution if and only if

$$\langle \rho^{(1)} \rangle \big|_{r=0} = 0, \tag{121}$$

$$\hat{L} \langle \rho^{(1)} \rangle - \frac{1}{2} \left( \frac{\partial \rho_1^{(1)}}{\partial z} \bigg|_{z=1} - {}_i^2 D \frac{\partial \rho_2^{(1)}}{\partial z} \bigg|_{z=-1} \right) = 0, \tag{122}$$

that is, when in the averaged problem for remainder term there are no sources. Averaging (107) with regard to (109), one can show that (122) is satisfied identically. Thus, in order to the averaged problem for the remainder term had a trivial solution it is sufficient for the average condition (121) to be satisfied. Therefore, in order to get the exact on the average solution of (44) - (50) on the field of density in the layer, in the formulation of the problem for the first coefficients of the asymptotic expansion (90) (91), (105) (95) - (97) the boundary condition (97) must be replaced by non-local (121).

## 5. Solution to the problem of mass transfer in the zero approximation

### 5.1. Solution to the problem in the zero approximation

In the image space of Laplace-Carson the problem (76) - (82) in the zero approximation is represented as

$$p\rho_1^{(0)u} - \frac{\partial^2 \rho_1^{(0)u}}{\partial z^2} = -At\,\rho_1^{(0)u}, \quad r>0,\ z>1, \tag{123}$$

$$p\rho_2^{(0)u} - {}_1^2D\frac{\partial^2 \rho_2^{(0)u}}{\partial z^2} = -At\,\rho_2^{(0)u}, \quad r>0,\ z<-1, \tag{124}$$

$$p\rho^{(0)u} + \frac{Pd}{r}\frac{\partial\rho^{(0)u}}{\partial r} + At\,\rho^{(0)u} = \frac{1}{2}\left(\left.\frac{\partial\rho_1^{(0)u}}{\partial z}\right|_{z=1} - {}_1^2D\left.\frac{\partial\rho_2^{(0)u}}{\partial z}\right|_{z=-1}\right), \quad r>0,\ |z|<1, \tag{125}$$

$$\left.\rho_1^{(0)u}\right|_{z=1} = \rho^{(0)u} = \left.\rho_2^{(0)u}\right|_{z=-1}, \tag{126}$$

$$\left.\rho^{(0)u}\right|_{r=0} = 1/\gamma, \tag{127}$$

$$\left.\rho^{(0)u}\right|_{r\to\infty} = 0, \quad \left.\rho_1^{(0)u}\right|_{r+z\to\infty} = 0, \quad \left.\rho_2^{(0)u}\right|_{r+|z|\to\infty} = 0. \tag{128}$$

Taking into account the boundary conditions (126), as well as the fact that in the zero approximation the density of the pollutant in the porous layer is independent of $z$ and is a function only of $r$ and $t$, the solution of equations (123), (124) can be rewritten as follows:

$$\rho_1^{(0)u} = \rho^{(0)u} \exp\left(-\sqrt{p+At}\,(z-1)\right), \tag{129}$$

$$\rho_2^{(0)u} = \rho^{(0)u} \exp\left(\sqrt{\tfrac{1}{2}D(p+At)}\,(z+1)\right). \tag{130}$$

These expressions allow us to determine the values of the traces of derivatives from the outer regions included in the equation for the layer, through the density of impurities in it

$$\left.\frac{\partial \rho_1^{(0)u}}{\partial z}\right|_{z=1} = -\sqrt{p+At}\,\rho^{(0)u}, \quad \left.\frac{\partial \rho_2^{(0)u}}{\partial z}\right|_{z=-1} = \sqrt{\tfrac{1}{2}D(p+At)}\,\rho^{(0)u}. \tag{131}$$

Substituting (131) into equation (125), after simple transformations we obtain an ordinary differential equation for the determination of $\rho^{(0)u}$

$$\frac{Pd}{r}\frac{d\rho^{(0)u}}{dr} = -\left(p+At+\frac{1}{2}\sqrt{p+At}\left(1+\sqrt{\tfrac{2}{i}D}\right)\right)\rho^{(0)u}, \tag{132}$$

from which we finally get

$$\rho^{(0)u} = \gamma \exp\left[-\left(p+At+\frac{1}{2}\sqrt{p+At}\left(1+\sqrt{\tfrac{2}{i}D}\right)\right)\frac{r^2}{2Pd}\right]. \tag{133}$$

Translation in the original space is carried out by the reference [51]. The expression for the density of radioactive contaminants for the porous layer in the original space is represented as

$$\rho^{(0)} = \frac{\gamma}{2}\exp\left(-\Lambda tr^2/2Pd\right)\Phi\left(t-r^2/2Pd\right)\times$$

$$\times\left[\exp\left(\frac{-\sqrt{At}\left(1+\sqrt{\tfrac{2}{i}D}\right)r^2}{4Pd}\right)\mathrm{erfc}\left(\frac{\sqrt{\delta}\left(1+\sqrt{\tfrac{2}{i}D}\right)r^2}{8Pd\sqrt{t-r^2/2Pd}}-\sqrt{At\left(t-r^2/2Pd\right)}\right)+\right.$$

$$\left.+\exp\left(\frac{\sqrt{At}\left(1+\sqrt{\tfrac{2}{i}D}\right)r^2}{4Pd}\right)\mathrm{erfc}\left(\frac{\left(1+\sqrt{\tfrac{2}{i}D}\right)r^2}{8Pd\sqrt{t-r^2/2Pd}}+\sqrt{At\left(t-r^2/2Pd\right)}\right)\right], \, t>0, \, r>0, \, |z|<1. \tag{134}$$

Also transition is feasible in the original space for the coating (129) and underlying (130) layers

$$
\rho_1^{(0)} = \frac{\gamma}{2}\exp\left(-Atr^2/2Pd\right)\Phi\left(t-r^2/2Pd\right)\times
$$
$$
\times\left\{\exp\left(-\sqrt{At}\left(1+\sqrt{{}_1^2D}\right)r^2\Big/4Pd-\sqrt{At}\left(z-1\right)\right)\times\right.
$$
$$
\times\,\mathrm{erfc}\left(\frac{\left(1+\sqrt{{}_1^2D}\right)r^2+4Pd(z-1)}{8Pd\sqrt{t-r^2/2Pd}}-\sqrt{At\left(t-\frac{r^2}{2Pd}\right)}\right)+
$$

$$
+\exp\left(\sqrt{At}\left(1+\sqrt{{}_1^2D}\right)r^2\Big/4Pd+\sqrt{At}\left(z-1\right)\right)\times
$$
$$
\left.\times\,\mathrm{erfc}\left(\frac{\left(1+\sqrt{{}_1^2D}\right)r^2+4Pd(z-1)}{8Pd\sqrt{t-r^2/2Pd}}+\sqrt{At\left(t-\frac{r^2}{2Pd}\right)}\right)\right\},\quad t>0,\ r>0,\ z>1,
$$

(135)

$$
\rho_2^{(0)} = \frac{\gamma}{2}\exp\left(-Atr^2/2Pd\right)\Phi\left(t-r^2/2Pd\right)\times\left\{\exp\left(-\sqrt{At}\left(1+\sqrt{{}_1^2D}\right)r^2\Big/4Pd+\sqrt{At\,{}_2^1D}\left(z+1\right)\right)\times\right.
$$
$$
\times\,\mathrm{erfc}\left(\frac{\left(1+\sqrt{{}_1^2D}\right)r^2-4Pd\sqrt{{}_2^1D}\left(z+1\right)}{8Pd\sqrt{t-r^2/2Pd}}-\sqrt{At\left(t-\frac{r^2}{2Pd}\right)}\right)+
$$

$$
+\exp\left(\sqrt{At}\left(1+\sqrt{{}_1^2D}\right)r^2\Big/4Pd-\sqrt{At_2^1 D}\left(z+1\right)\right)\times
$$
$$
\left.\times\,\mathrm{erfc}\left(\frac{\left(1+\sqrt{{}_1^2D}\right)r^2-4Pd\sqrt{{}_2^1D}\left(z+1\right)}{8Pd\sqrt{t-r^2/2\gamma Pt}}+\sqrt{At\left(t-\frac{r^2}{2Pd}\right)}\right)\right\},\ t>0,\ r>0,\ z<-1.
$$

(136)

The first factor in the solution (134) - (136) describes the decrease in the density of the pollutant as a result of radioactive decay, the second - the Heaviside function, determines the radius of the spread of contamination zone and the third (the expression in curly brackets) takes into account changes of the density due to the diffusion of pollutants, and radioactive decay of diffusing nuclide. Since the contribution of radioactive decay is described by the factor of $\exp\left(-Atr^2/2Pd\right)$, then it can be argued that the concentration of the radioactive contaminant is reduced by a factor of $e$ due to the decay at the distances defined by the simple relation $R_e=h\sqrt{2Pd/At}=\sqrt{2v_0r_0/\alpha}$. It follows that for short-lived isotopes, zone of contamination is low. On the other hand, to reduce the zone of influence of long-lived radioactive isotopes, the rate of filtration should be reduced.

The resulting solution (134) contains the Heaviside function, which vanishes for $r \geq \sqrt{2Pd\,t}$, and helps to determine the radius of the zone of radioactive contamination

$$R_p = h\sqrt{2\mathrm{Pd}\,t} = \sqrt{2v_0 r_0 \tau}. \tag{137}$$

The most important physical results are described by the zero approximation of the asymptotic expansion, the first and the following coefficients determine the "correction term". In addition, due to the smallness of the diffusion coefficient ($D_z \sim 10^{-9} \div 10^{-11}$), the spread of a contaminant in water-resistant layers in the vertical direction is negligible compared to the convective transport in a porous layer and has little effect on the size of the zone of contamination.

## 5.2. The solution to the problem of mass transfer in the first approximation

In the space transformations of Laplace-Carson, the problem (90), (91), (105), (95) - (97), (121) for the first coefficient of expansion is represented as

$$\hat{L}^u \rho^{(1)u} = \frac{{}_0 D}{4}\left(z^2 - \frac{1}{3}\right)\hat{L}^u\left(\left.\frac{\partial \rho_1^{(0)u}}{\partial z}\right|_{z=1} - {}_1^2 D\left.\frac{\partial \rho_2^{(0)u}}{\partial z}\right|_{z=-1}\right) +$$
$$+ \frac{{}_0 D}{2}z\hat{L}^u\left(\left.\frac{\partial \rho_1^{(0)u}}{\partial z}\right|_{z=1} + {}_1^2 D\left.\frac{\partial \rho_2^{(0)u}}{\partial z}\right|_{z=-1}\right) + \frac{1}{2}\left(\left.\frac{\partial \rho_1^{(1)u}}{\partial z}\right|_{z=1} - {}_1^2 D\left.\frac{\partial \rho_2^{(1)u}}{\partial z}\right|_{z=-1}\right), \quad r > 0,\ |z| < 1, \tag{138}$$

$$p\rho_1^{(1)u} - \frac{\partial^2 \rho_1^{(1)u}}{\partial z^2} + \mathrm{At}\,\rho_1^{(1)u} = 0, \quad r > 0,\ z > 1, \tag{139}$$

$$p\rho_2^{(1)u} - {}_1^2 D\frac{\partial^2 \rho_2^{(1)u}}{\partial z^2} + \mathrm{At}\,\rho_2^{(1)u} = 0, \quad r > 0,\ z < -1. \tag{140}$$

The initial conditions and conjugation conditions at the boundaries are represented as

$$\left.\rho^{(1)u}\right|_{z=1} = \left.\rho_1^{(1)u}\right|_{z=1}, \tag{141}$$

$$\left.\rho^{(1)u}\right|_{z=-1} = \left.\rho_2^{(1)u}\right|_{z=-1}, \tag{142}$$

$$\left.\rho^{(1)u}\right|_{r\to+\infty} = 0,\ \left.\rho_1^{(1)u}\right|_{r+z\to+\infty} = 0,\ \left.\rho_2^{(1)u}\right|_{r+|z|\to+\infty} = 0, \tag{143}$$

$$\left\langle \rho^{(1)u} \right\rangle \Big|_{r=0} = 0. \tag{144}$$

The operator $\hat{L}$ in the image space has the form

$$\hat{L}^{\,u} = p + At + \gamma \frac{Pt}{r} \frac{\partial}{\partial r}.$$

The action of this operator on the zero expansion in the image space is determined by the formula

$$\hat{L}^{u} \rho^{(0)u} = -\frac{\sqrt{p+At}}{2}\left(1+\sqrt{\tfrac{2}{i}D}\right)\rho^{(0)u}. \tag{145}$$

The solution in the first approximation, according to (71), is sought in the form of quadratic polynomial

$$\rho^{(1)u} = {}^{1}_{0}D\left(\frac{z^2}{2}E^u + z F^u + Q^u\right), \tag{146}$$

in which $E^{\,u}$ and $F^{\,u}$ are expressed through the zero approximation according to (73), (74)

$$E^{u} = \frac{1}{2}\left(\frac{\partial \rho_1^{(0)u}}{\partial z}\bigg|_{z=1} - {}^{2}_{1}D \frac{\partial \rho_2^{(0)u}}{\partial z}\bigg|_{z=-1}\right), F^{u} = \frac{1}{2}\left(\frac{\partial \rho_1^{(0)u}}{\partial z}\bigg|_{z=1} + {}^{2}_{1}D \frac{\partial \rho_2^{(0)u}}{\partial z}\bigg|_{z=-1}\right), \tag{147}$$

and the function $Q^{\,u}$, according to (104), is defined by the equation

$$\hat{L}^{u} Q^{u} = \frac{{}^{0}_{1}D}{2}\left(\frac{\partial \rho_1^{(1)u}}{\partial z}\bigg|_{z=1} - {}^{2}_{1}D \frac{\partial \rho_2^{(1)u}}{\partial z}\bigg|_{z=-1}\right) - \frac{1}{6}\hat{L}^{u} E^{u}. \tag{148}$$

Solutions of equations (139), (140) shall be as follows:

$$\rho_1^{(1)u} = \rho^{(1)u}\bigg|_{z=1} \exp\left(-\sqrt{p+At}\,(z-1)\right), \rho_2^{(1)u} = \rho^{(1)u}\bigg|_{z=-1} \exp\left(\sqrt{\tfrac{1}{2}D(p+At)}\,(z+1)\right). \tag{149}$$

Let us find the traces of the outer regions of the right side of equation (138)

$$\left.\frac{\partial \rho_1^{(1)u}}{\partial z}\right|_{z=1} = -\,_0^1 D \sqrt{p+At}\left(\frac{1}{2}E^u + F^u + Q^u\right),$$

(150)

$$\left.\,_1^2 D \frac{\partial \rho_2^{(1)u}}{\partial z}\right|_{z=-1} = \,_0^2 D \sqrt{\,_1^2 D(p+At)}\left(\frac{1}{2}E^u - F^u + Q^u\right).$$

(151)

Note also that the action of the operator $\hat{L}^{\,u}$ on functions

$$E^u = -\frac{\sqrt{p+At}}{2}\left(1+\sqrt{\,_1^2 D}\right)\rho^{(0)u}, F^u = -\frac{\sqrt{\delta(p+At)}}{2}\left(1-\sqrt{\,_1^2 D}\right)\rho^{(0)u}$$

(152)

according to (147), this leads to the following:

$$\hat{L}^u E^u = \frac{p+At}{4}\left(1+\sqrt{\,_1^2 D}\right)^2 \rho^{(0)u}, \hat{L}^u F^u = \frac{p+At}{4}\left(1-\,_1^2 D\right)^2 \rho^{(0)u}.$$

(153)

The final equation for determination of $Q^u$ takes the form

$$\frac{Pd}{r}\frac{dQ^u}{dr}+\left(p+At+\frac{\sqrt{p+At}}{2}\left(1+\sqrt{\,_1^2 D}\right)\right)Q^u = \frac{1+\,_1^2 D-\sqrt{\,_1^2 D}}{3}(p+At)\rho^{(0)u}.$$

(154)

It's general solution is represented as

$$Q^u = \frac{1+\,_1^2 D-\sqrt{\,_1^2 D}}{3Pd}\delta(p+At)\times\int_0^r \rho^{(0)u}\exp\left[-\left(p+At+\frac{\sqrt{p+At}}{2}\left(1+\sqrt{\,_1^2 D}\right)\right)\frac{r^2-r'^2}{2Pd}\right]r'dr'\,+$$

$$+\,C\exp\left[-\left(p+At+\frac{\sqrt{p+At}}{2}\left(1+\sqrt{\,_1^2 D}\right)\right)\frac{r^2}{2Pd}\right].$$

(155)

Constant $C$ is determined from the averaging (144)

$$Q^u\Big|_{r=0} = -\frac{1}{6}E^u\Big|_{r=0}. \tag{156}$$

Hence, one can obtain

$$Q^u = \frac{1}{6}\left[\frac{\left(1+{}_1^2D-\sqrt{{}_1^2D}\right)}{Pd}\delta(p+At)r^2 + \frac{\sqrt{p+At}}{2}\left(1+\sqrt{{}_1^2D}\right)\right]\rho^{(0)u}. \tag{157}$$

As a result, the solution for the first coefficient of expansion in the images is represented as

$$\rho^{(1)u} = \frac{{}_0^1D}{2}\sqrt{p+At}\left[\left(\frac{1}{6}-\frac{z^2}{2}\right)\left(1+\sqrt{{}_1^2D}\right)-z\left(1-\sqrt{{}_1^2D}\right)\right]\rho^{(0)u}+{}_0^1D\frac{1+{}_1^2D-\sqrt{{}_1^2D}}{6Pd}r^2(p+At)\rho^{(0)u}, \tag{158}$$

$$\rho_1^{(1)u} = \frac{{}_0^1D}{3}\left[\frac{1+{}_1^2D-\sqrt{{}_1^2D}}{2Pd}r^2(p+At)-\sqrt{p+At}\left(2-\sqrt{{}_1^2D}\right)\right]\rho^{(0)u}\times\exp\left(-(z-1)\sqrt{p+At}\right), \tag{159}$$

$$\rho_2^{(1)u} = \frac{{}_0^1D}{3}\left[\frac{1+{}_1^2D-\sqrt{{}_1^2D}}{2Pd}r^2(p+At)+\sqrt{p+At}\left(1-2\sqrt{{}_1^2D}\right)\right]\rho^{(0)u}\times\exp\left((z+1)\sqrt{{}_2^1D(p+At)}\right). \tag{160}$$

Determination of the originals is carried out by help of the following correspondence:

$$\sqrt{p+\gamma}\exp\left(-\sqrt{\beta(p+\gamma)}\right)\rightarrow\frac{1}{\sqrt{\pi}\,t}\exp\left(-\gamma t-\frac{\beta}{4t}\right)-$$

$$-\frac{\sqrt{\gamma}}{2}\left[\exp(\sqrt{\beta\gamma})\,\mathrm{erfc}\left(\frac{1}{2}\sqrt{\frac{\beta}{t}}+\sqrt{\gamma t}\right)-\exp(-\sqrt{\beta\gamma})\,\mathrm{erfc}\left(\frac{1}{2}\sqrt{\frac{\beta}{t}}-\sqrt{\gamma t}\right)\right],$$

$$p\exp\left(-\sqrt{\beta(p+\gamma)}\right)\rightarrow\frac{\sqrt{\beta}}{2t\sqrt{\pi t}}\exp\left(-\frac{\beta}{4t}-\gamma t\right).$$

Finally, we obtain for the porous layer

$$\rho^{(1)} = \frac{\frac{1}{0}D}{4\gamma}\sqrt{At}\exp\left(-Atr^2/2Pd\right)\Phi\left(t-r^2/2Pd\right)\times$$

$$\times\left[\frac{2\left(1+\sqrt{\frac{2}{1}D}\right)\sqrt{At\left(t-r^2/2Pd\right)}}{\sqrt{\pi}}\left(\frac{1-\sqrt{\frac{2}{1}D}+\frac{2}{1}D}{24Pd^2\left(t-r^2/2Pd\right)}r^4+\left(\frac{1}{6}-\frac{z^2}{2}-z\frac{1-\sqrt{\frac{2}{1}D}}{1+\sqrt{\frac{2}{1}D}}\right)\right)\times\right.$$

$$\times\exp\left(-\frac{\left(1+\sqrt{\frac{2}{1}D}\right)^2}{64Pd^2\left(t-r^2/2Pd\right)}r^4-At\left(t-\frac{r^2}{2Pd}\right)\right)+$$

$$+\left(\sqrt{At}\frac{1-\sqrt{\frac{2}{1}D}+\frac{2}{1}D}{3Pd}r^2-\left(1+\sqrt{\frac{2}{1}D}\right)\left(\frac{1}{6}-\frac{z^2}{2}-z\frac{1-\sqrt{\frac{2}{1}D}}{1+\sqrt{\frac{2}{1}D}}\right)\right)\times$$

$$\times\exp\left(\frac{\sqrt{At}\left(1+\sqrt{\frac{2}{1}D}\right)}{4Pd}r^2\right)\mathrm{crfc}\left(\frac{1+\sqrt{\frac{2}{1}D}}{8Pd\sqrt{t-r^2/2Pd}}r^2+\sqrt{At\left(t-\frac{r^2}{2Pd}\right)}\right)+$$

$$+\left(\sqrt{At}\frac{1-\sqrt{\frac{2}{1}D}+\frac{2}{1}D}{3Pd}r^2+\left(1+\sqrt{\frac{2}{1}D}\right)\left(\frac{1}{6}-\frac{z^2}{2}-z\frac{1-\sqrt{\frac{2}{1}D}}{1+\sqrt{\frac{2}{1}D}}\right)\right)\times$$

$$\times\exp\left(-\frac{\sqrt{At}\left(1+\sqrt{\frac{2}{1}D}\right)}{4Pd}r^2\right)\mathrm{erfc}\left(\frac{1+\sqrt{\frac{2}{1}D}}{8Pd\sqrt{t-r^2/2Pd}}r^2-\sqrt{At\left(t-\frac{r^2}{2Pd}\right)}\right)\right], \quad r>0,\ t>0,\ |z|<1, \tag{161}$$

for the covering layer

$$\rho_1^{(1)} = \frac{1}{0}D\frac{\sqrt{At}}{6\gamma}\exp\left(\frac{At}{2Pd}r^2\right)\Phi\left(t-\frac{r^2}{2Pd}\right)\times$$

$$\times\left[\frac{2}{\sqrt{\pi}\sqrt{At\left(t-r^2/2Pd\right)}}\left(\frac{1+\frac{2}{1}D-\sqrt{\frac{2}{1}D}}{4Pd\left(t-r^2/2Pd\right)}r^2\left(\frac{1+\sqrt{\frac{2}{1}D}}{4Pd}r^2+z-1\right)-\left(2-\sqrt{\frac{2}{1}D}\right)\right)\times\right.$$

$$\times\exp\left(-\frac{\left(\left(1+\sqrt{\frac{2}{1}D}\right)r^2+4Pd(z-1)\right)^2}{64Pd^2\left(t-r^2/2Pd\right)}-At\left(t-\frac{r^2}{2Pd}\right)\right)+$$

$$+\left(\frac{\sqrt{At}\left(1+\frac{2}{1}D-\sqrt{\frac{2}{1}D}\right)}{2Pd}r^2-\left(2-\sqrt{\frac{2}{1}D}\right)\right)\exp\left(-\frac{\sqrt{At}\left(1+\sqrt{\frac{2}{1}D}\right)}{4Pd}r^2-\sqrt{At}(z-1)\right)\times$$

$$\times\mathrm{erfc}\left(\frac{\left(1+\sqrt{\frac{2}{1}D}\right)r^2}{8Pd\sqrt{t-r^2/2Pd}}+\frac{z-1}{2\sqrt{t-r^2/2Pd}}-\sqrt{At\left(t-\frac{r^2}{2Pd}\right)}\right)+$$

$$+\left(\frac{\sqrt{At}\left(1+\frac{2}{1}D-\sqrt{\frac{2}{1}D}\right)}{2Pd}r^2+\left(2-\sqrt{\frac{2}{1}D}\right)\right)\exp\left(-\frac{\sqrt{At}\left(1+\sqrt{\frac{2}{1}D}\right)}{4Pd}r^2+\sqrt{At}(z-1)\right)\times$$

$$\times\mathrm{erfc}\left(\frac{\left(1+\sqrt{\frac{2}{1}D}\right)r^2}{8Pd\sqrt{t-r^2/2Pd}}+\frac{z-1}{2\sqrt{t-r^2/2Pd}}+\sqrt{At\left(t-\frac{r^2}{2Pd}\right)}\right)\right]. \tag{162}$$

and for the underlying layer

$$
\rho_2^{(1)} = {}_0^1 D \frac{\sqrt{At}}{6\gamma} \exp\left(-\frac{At}{2Pd}r^2\right)\Phi\left(t-\frac{r^2}{2Pd}\right) \times \times \left[\frac{2}{\sqrt{\pi}\sqrt{At\left(t-r^2/2Pd\right)}}\left(\frac{1+{}_1^2D-\sqrt{{}_1^2D}}{4Pd\left(t-r^2/2Pd\right)}r^2\left(\frac{1+\sqrt{{}_1^2D}}{4Pd}r^2-\sqrt{{}_2^1D}(z+1)\right)+1-2\sqrt{{}_1^2D}\right) \times \right.
$$

$$
\times \exp\left(-\frac{\left(\left(1+\sqrt{{}_1^2D}\right)r^2-4Pd\sqrt{{}_2^1D}(z+1)\right)^2}{64Pd^2\left(t-r^2/2Pd\right)} - At\left(t-\frac{r^2}{2Pd}\right)\right)+
$$

$$
+\left(\frac{\sqrt{At}\left(1+{}_1^2D-\sqrt{{}_1^2D}\right)}{2Pd}r^2+1-2\sqrt{{}_1^2D}\right)\exp\left(-\frac{\sqrt{At}\left(1+\sqrt{{}_1^2D}\right)}{4Pd}r^2+\sqrt{{}_2^1D At}\,(z+1)\right) \times
$$

$$
\times \mathrm{erfc}\left(\frac{\left(1+\sqrt{{}_1^2D}\right)r^2-4Pd\sqrt{{}_2^1D}(z+1)}{8Pd\sqrt{t-r^2/2Pd}}-\sqrt{At\left(t-\frac{r^2}{2Pd}\right)}\right)+
$$

$$
+\left(\frac{\sqrt{At}\left(1+{}_1^2D-\sqrt{{}_1^2D}\right)}{2Pd}r^2-1+2\sqrt{{}_1^2D}\right)\exp\left(-\frac{\sqrt{At}\left(1+\sqrt{{}_1^2D}\right)}{4Pd}r^2-\sqrt{{}_2^1D At}\,(z+1)\right) \times
$$

$$
\left.\times \mathrm{erfc}\left(\frac{\left(1+\sqrt{{}_1^2D}\right)r^2-4Pd\sqrt{{}_2^1D}(z+1)}{8Pd\sqrt{t-r^2/2Pd}}+\sqrt{At\left(t-\frac{r^2}{2Pd}\right)}\right)\right], \quad r>0,\ t>0,\ z<-1. \tag{163}
$$

Note that when $r = 0$ the first coefficient of expansion (158)

$$
\rho^{(1)u}\Big|_{r=0} = \frac{{}_0^1 D}{2}\sqrt{p+At}\left[\frac{1}{2}\left(\frac{1}{3}-z^2\right)\left(1+\sqrt{{}_1^2D}\right)-z\left(1-\sqrt{{}_1^2D}\right)\right] \tag{164}
$$

depends on the variable $z$ and does not vanish, that is constructed solution does not satisfy the boundary condition (48). To eliminate the marked lack, it is necessary to build a boundary-layer functions in a neighborhood of $r = 0$.

## 5.3. Construction of boundary-layer solutions

Let us represent the solution to (78) - (82) as

$$
\rho_1 = \hat{\rho}_1 + \Pi_1, \rho = \hat{\rho} + \Pi, \rho_2 = \hat{\rho}_2 + \Pi_2 \tag{165}
$$

where $\hat{\rho} = \hat{\rho}(r, z, t)$ - the regular part, $\Pi = \Pi(y, z, t)$ - boundary-layer part of the expansion by the asymptotic parameter, $y = r^2/2\varepsilon$ the stretched variable [44]. Substituting (165) to (44) - (50) and applying the transformation of Laplace-Carson over the variable $t$, in the image space we obtain a problem for the boundary-layer functions

$$
\Pi^{\hat{e}} = \Pi^{(0)u} + \varepsilon\Pi^{(1)u} + ..., \Pi_1^u = \Pi_1^{(0)u} + \varepsilon\Pi_1^{(1)u} + ..., \Pi_2^u = \Pi_2^{(0)u} + \varepsilon\Pi_2^{(1)u} + ... \tag{166}
$$

in the following way:

$$\left( \frac{\partial^2 \Pi_1^{(0)u}}{\partial z^2} - (p + At)\Pi_1^{(0)u} \right) + \varepsilon \left( \frac{\partial^2 \Pi_1^{(1)u}}{\partial z^2} - (p + At)\Pi_1^{(1)u} \right) + \ldots = 0, \quad y > 0, \quad z > 1, \tag{167}$$

$$\left( \frac{\partial^2 \Pi^{(0)u}}{\partial z^2} - \frac{1}{0} DPd \frac{\partial \Pi^{(0)u}}{\partial y} \right) + \varepsilon \left( \frac{\partial^2 \Pi^{(1)u}}{\partial z^2} - \frac{1}{0} DPd \frac{\partial \Pi^{(1)u}}{\partial y} - (p + At)\Pi^{(0)u} \right) + \ldots = 0, \quad y > 0, |z| < 1, \tag{168}$$

$$\left( \frac{\partial^2 \Pi_2^{(0)u}}{\partial z^2} - \frac{1}{2} D(p + At)\Pi_2^{(0)u} \right) + \varepsilon \left( \frac{\partial^2 \Pi_2^{(1)u}}{\partial z^2} - \frac{1}{2} D(p + At)\Pi_2^{(1)u} \right) + \ldots = 0, \quad y > 0, \quad z < -1, \tag{169}$$

$$\frac{\partial \Pi^{(0)u}}{\partial z}\Big|_{z=1} + \varepsilon \left( \frac{\partial \Pi^{(1)u}}{\partial z}\Big|_{z=1} - \frac{1}{0} D \frac{\partial \Pi_1^{(0)u}}{\partial z}\Big|_{z=1} \right) + \ldots = 0, \quad \frac{\partial \Pi^{(0)u}}{\partial z}\Big|_{z=-1} + \varepsilon \left( \frac{\partial \Pi^{(1)u}}{\partial z}\Big|_{z=-1} - \frac{2}{0} D \frac{\partial \Pi_2^{(0)u}}{\partial z}\Big|_{z=-1} \right) + \ldots = 0, \tag{170}$$

$$\left( \Pi^{(0)\grave{e}} - \Pi_1^{(0)u} + \varepsilon \left( \Pi^{(1)u} - \Pi_1^{(1)u} \right) + \ldots \right)\Big|_{z=1} = 0, \left( \Pi^{(0)u} - \Pi_2^{(0)u} + \varepsilon \left( \Pi^{(1)u} - \Pi_2^{(1)u} \right) + \ldots \right)\Big|_{z=-1} = 0, \tag{171}$$

$$\Pi^{(0)u}\Big|_{y=0} + \varepsilon \left( \rho^{(1)u} + \Pi^{(1)u} \right)\Big|_{y=0} + \ldots = 0, \tag{172}$$

$$\left( \Pi^{(0)u} + \varepsilon \Pi^{(1)u} + \ldots \right)\Big|_{y\to\infty} = 0, \left( \Pi_1^{(0)u} + \varepsilon \Pi_1^{(1)u} + \ldots \right)\Big|_{y+z\to\infty} = 0, \left( \Pi_2^{(0)u} + \varepsilon \Pi_2^{(1)u} + \ldots \right)\Big|_{y+|z|\to\infty} = 0. \tag{173}$$

The problem for the zero coefficients has only the trivial solution

$$\Pi^{(0)u} = \Pi_1^{(0)u} = \Pi_2^{(0)u} = 0. \tag{174}$$

The problem for the first coefficient is divided into three independent parts. The first one is the problem for the boundary layer functions in the layer of the form

$$\frac{\partial^2 \Pi^{(1)u}}{\partial z^2} - (p + At)\frac{\partial \Pi^{(1)u}}{\partial y} = 0, y > 0, \quad |z| < 1, \tag{175}$$

$$\frac{\partial \Pi^{(1)u}}{\partial z}\bigg|_{z=1} = \frac{\partial \Pi^{(1)u}}{\partial z}\bigg|_{z=-1} = 0, \tag{176}$$

$$\Pi^{(1)u}\bigg|_{y=0} = -\rho^{(1)u}\bigg|_{y=0} = -\sqrt{p + At}\,\frac{{}^1_0 D}{2}\left[\left(\frac{1}{6} - \frac{z^2}{2}\right)\left(1 + \sqrt{{}^2_1 D}\right) - z\left(1 - \sqrt{{}^2_1 D}\right)\right], \tag{177}$$

$$\Pi^{(1)u}\bigg|_{y\to\infty} = 0. \tag{178}$$

The solution to this problem found by the method of separation of variables and is defined by the formula

$$\Pi^{(1)u} = \frac{4\sqrt{p+At}\;{}^0_1 D}{\pi^2}\left(1 - \sqrt{{}^2_1 D}\right)\sin\left(\frac{\pi}{2}z\right)\exp\left(-\left(\frac{\pi}{2}\right)^2\frac{{}^0_1 D}{Pd}y\right) +$$

$$+ \frac{\sqrt{p+At}\;{}^0_1 D}{\pi^2}\sum_{n=1}^{\infty}\left(4\frac{(-1)^n}{(1+2n)^2}\left(1 - \sqrt{{}^2_1 D}\right)\sin\left[\left(\frac{\pi}{2} + \pi n\right)z\right]\exp\left(-\left(\frac{\pi}{2} + \pi n\right)^2\frac{{}^0_1 D}{Pd}y\right) + \tag{179}$$

$$+ \frac{(-1)^n}{n^2}\left(1 + \sqrt{{}^2_1 D}\right)\cos(\pi nz)\exp\left(-(\pi n)^2\frac{{}^0_1 D}{Pd}y\right)\right).$$

Returning to the original, we obtain an expression for the first coefficient of expansion of the boundary-layer functions in the layer

$$\Pi^{(1)} = \frac{{}^0_1 D}{\pi^2}\left(\frac{\exp(-t\cdot At)}{\sqrt{\pi t}} + \sqrt{At}\,erf\sqrt{t\cdot Ad}\right)\times\left\{4\left(1 - \sqrt{{}^2_1 D}\right)\sin\left(\frac{\pi}{2}z\right)\exp\left(-\left(\frac{\pi}{2}\right)^2\frac{{}^0_1 D}{Pd}y\right) +$$

$$+ \sum_{n=1}^{\infty}\left(4\frac{(-1)^n}{(1+2n)^2}\left(1 - \sqrt{{}^2_1 D}\right)\sin\left[\left(\frac{\pi}{2} + \pi n\right)z\right]\exp\left(-\left(\frac{\pi}{2} + \pi n\right)^2\frac{{}^0_1 D}{Pd}y\right) + \tag{180}$$

$$+ \frac{(-1)^n}{n^2}\left(1 + \sqrt{{}^2_1 D}\right)\cos(\pi nz)\exp\left(-(\pi n)^2\frac{{}^0_1 D}{Pd}y\right)\right)\right\}.$$

Equation (180) allows determining the boundary values of the first boundary layer coefficients for the surrounding half-spaces, the problem to determine which are the following:

$$\frac{\partial^2 \Pi_1^{(1)u}}{\partial z^2} - (p + At)\Pi_1^{(1)u} = 0, z > 1, \Pi_1^{(1)u}\Big|_{z=1} = \Pi^{(1)u}\Big|_{z=1}, \Pi_1^{(1)u}\Big|_{y+z\to\infty} = 0, \tag{181}$$

$$\frac{\partial^2 \Pi_2^{(1)u}}{\partial z^2} - \tfrac{1}{2}D(p + At)\Pi_2^{(1)u} = 0, z < -1, \Pi_2^{(1)u}\Big|_{z=-1} = \Pi^{(1)u}\Big|_{z=-1}, \Pi_2^{(1)u}\Big|_{y+|z|\to+\infty} = 0. \tag{182}$$

Solutions to (181), (182) is represented through $\Pi^{(1)u}$ by the following form:

$$\Pi_1^{(1)u} = \Pi^{(1)u}\Big|_{z=1} \exp\left(-\sqrt{p + At}\,(z-1)\right) \tag{183}$$

$$\Pi_2^{(1)u} = \Pi^{(1)u}\Big|_{z=-1} \exp\left(\sqrt{\tfrac{1}{2}D(p + At)}\,(z+1)\right) \tag{184}$$

which in the originals are represented as

$$
\begin{aligned}
\Pi_1^{(1)} = & \left\{ \frac{1}{\sqrt{\pi t}} \exp\left(-t\cdot At - \frac{(z-1)^2}{4t}\right) - \frac{\sqrt{At}}{2}\left[ \exp\left[(z-1)\sqrt{At}\right]\mathrm{erfc}\left(\frac{z-1}{2\sqrt{t}} + \sqrt{t\cdot At}\right) - \right.\right. \\
& \left.\left. -\exp\left[-(z-1)\sqrt{At}\right]\mathrm{erfc}\left(\frac{z-1}{2\sqrt{t}} - \sqrt{t\cdot At}\right)\right]\right\} \left(\frac{4\,{}^0_1D}{\pi^2}\left(1 - \sqrt{{}^2_1D}\right)\exp\left[-\left(\frac{\pi}{2}\right)^2 \frac{{}^0_1D}{\mathrm{Pd}}y\right] + \\
& + \frac{{}^0_1D}{\pi^2}\sum_{n=1}^{\infty}\left(\frac{4(-1)^{2n}}{(1+2n)^2}\left(1 - \sqrt{{}^2_1D}\right)\exp\left[-\left(\frac{\pi}{2} + \pi n\right)^2 \frac{{}^0_1D}{\mathrm{Pd}}y\right] + \\
& + \frac{(-1)^{2n}}{n^2}\left(1 + \sqrt{{}^2_1D}\right)\exp\left[-(\pi n)^2 \frac{{}^0_1D}{\mathrm{Pd}}y\right]\right),
\end{aligned}
\tag{185}
$$

$$\Pi_2^{(1)} = \left\{ \frac{1}{\sqrt{\pi t}} \exp\left( -t \cdot Ad - \frac{{}_1^2 D (z+1)^2}{4t} \right) - \right.$$

$$- \frac{\sqrt{At}}{2} \left[ \exp\left[ -(z+1)\sqrt{{}_1^2 D At} \right] \mathrm{erfc}\left( \sqrt{t \cdot At} - \frac{(z+1)\sqrt{{}_1^2 D}}{2\sqrt{t}} \right) - \right.$$

$$\left. \left. - \exp\left[ (z+1)\sqrt{{}_1^2 D At} \right] \mathrm{erfc}\left( -\frac{(z+1)\sqrt{{}_1^2 D}}{2\sqrt{t}} - \sqrt{t \cdot At} \right) \right] \right\} \times$$

$$\times \left( \frac{4 {}_1^0 D}{\pi^2} \left( \sqrt{{}_1^2 D} - 1 \right) \exp\left[ -\left( \frac{\pi}{2} \right)^2 \frac{{}_1^0 D}{\mathrm{Pd}} y \right] + \right.$$

$$+ \frac{{}_1^0 D}{\pi^2} \sum_{n=1}^{\infty} \left( \frac{4(-1)^{2n+1}}{(1+2n)^2} \left( 1 - \sqrt{{}_1^2 D} \right) \exp\left[ -\left( \frac{\pi}{2} + \pi n \right)^2 \frac{{}_1^0 D}{\mathrm{Pd}} y \right] + \right.$$

$$\left. \left. + \frac{(-1)^{2n}}{n^2} \left( 1 + \sqrt{{}_1^2 D} \right) \exp\left[ -(\pi n)^2 \frac{{}_1^0 D}{\mathrm{Pd}} y \right] \right] \right).$$

(186)

The solution to the nonstationary problem (44) - (50) in the asymptotic approximation, taking into account the boundary-layer functions, represented as the sum of (134), (161) and (180) in the layer and, relevantly, for the surrounding area

$$\rho = \rho^{(0)} + \varepsilon \left( \rho^{(1)} + \Pi^{(1)} \right), \rho_1 = \rho_1^{(0)} + \varepsilon \left( \rho_1^{(1)} + \Pi_1^{(1)} \right), \rho_2 = \rho_2^{(0)} + \varepsilon \left( \rho_2^{(1)} + \Pi_2^{(1)} \right).$$

(187)

## 6. Conclusion

Figure 2 shows the dependence (134), the density of radioactive contaminants on the radial coordinate in the zero approximation for different times of injection with (curves 1, 2, 3) and without (curve 4), radioactive decay, as well as in diffusional approximation (curve 5). In many technological and non-technological liquid wastes of atomic industry, as one of the component contains a radioactive isotope of strontium $Sr_{90}^{38}$ with a half-life $T_{1/2} = 28$ years, which is a very dangerous chemical element for the biological structures due to its ability to replace atoms of calcium. Calculated options: half-thickness of the layer $h = 1$m, the diffusion coefficients $D_{1z} = D_{2z} = 10^{-11}$m$^2$/c, $D_z = 10^{-9}$ m$^2$/c; borehall radius of 0.1 m, the polluter - Strontium 90; activity of the solution - 1 Ci / l (high level waste ), the initial density of radioactive substances in the solution 7.34 g/m$^3$; injection volume 100 m$^3$/day (At = 75.8, Pe =1.84 $10^7$)

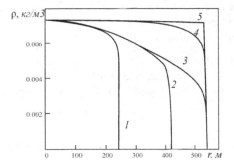

**Figure 2.** Plots of density on distance for different times of observation: Given the radioactive decay of *1* - 5 years, *2* - 15 years old, *3* - 25 years, *4* - 25 years (in the absence of decay At = 0), *5* - 25 years (in the absence of decay in non-diffusive approximation)

Figure 2 allows carrying out estimate of the contribution of diffusion and radioactive decay. The figure shows that the account of the radioactive decay (see curves 3 and 4) is the priority in comparison with the account of the mass transfer of the layer (curves 4 and 5) with the surrounding formations by diffusion. The figure also implies that the contribution of diffusion is essential in the front zone of radioactive contamination, where it is comparable with the magnitude of the density of matter in solution. With the zero approximation the size of the zone of contamination are determined.

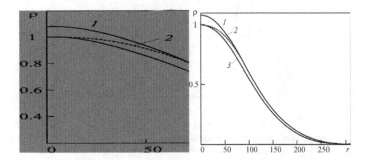

**Figure 3.** The dependence of the density of radioactive contaminants on the radial coordinate for the dimensionless time $t = 0.01$ (a), $t = 0.001$ (b): *1.2* - without and with taking into account the boundary-layer correction, respectively, *3* - zero approximation at $z = 0$

On the fig. 3 a, b it is shown that taking into account the boundary layer solution eliminates the disadvantage of the first approximation, which consists in the fact that the density of radioactive contaminants exceeds one (curve 1). In the calculations agreed: $Pd = 2 \cdot 10^7$, $^1_0D = 0.01$, $^2_1D = 1$, $\varepsilon = 0.05$, $Ad = 2200$, which corresponds to the half-life $T_{1/2} = 1$ year. Boundary layer correction results in the behavior of the curve in accordance with the conditions of the problem (see curves 1, 2) and at the same time clarifies the first approximation. Note that for short times accounting of the first expansion coefficient leads to refinements of the design parameters up to 20% (curves 1 and 3 in Fig.3 b). At the same time the importance of taking into account the boundary-layer functions in the near-well zone is illustrated ($r < 70$).

As the figure 3 b shows, on the border of the front distribution of contaminants occurs a sharp jump of density, indicating the presence of internal moving boundary layer, the patterns of which have not been investigated, and mathematical methods of eliminating it - not developed.

## Appendix: Note on the stationary solution to the problem

It is extremely important to find a stationary solution to set the maximum size of the zone of contamination. The equations describing the stationary regime are obtained from (44) - (50) if we consider the first term $(\partial \rho_i / \partial t)$ to be zero. The solution to this problem is given in [39, 40, 43]. Here we note that the solution to the stationary problem can be obtained from the first asymptotic approximation for $t \to \infty$.

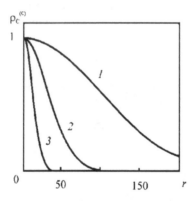

**Figure 4.** The dependence of the densities of radioactive contaminants in the porous layer for the stationary case (zero approximation) on the distance from the borehole at different decay constants: 1 - Ad = 0.01, 2 - 0.1, 3 - 1. Other parameters: Pd = $10^2$, $\delta = 10^{-4}$, $D_t^2 = 1$

Fig. 4 shows the dependence of the density of radioactive contaminants on the radial coordinate at the center of layer for various radioactive contaminants: curve 1 - $^{239}$Pu ($T_{1/2} = 2,24 \; 10^4$ years), curve 2 – $^{226}$Ra ($T_{1/2} = 1590$ years), curve 3 – $^{90}$Sr ($T_{1/2} = 28$ years).

The zero approximation in this case is the most important; it determines the general form of the dependence. The value of the density of the pollutant decreases exponentially, and as follows from the graphs, even for the middle half-life and most dangerous radionuclides ($^{90}$Sr, $^{137}$Cs) at distances of 200 $h$ (200 m) of the order of percent of maximum, observed in the area of injection.

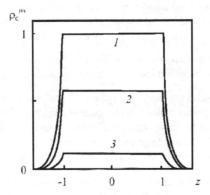

**Figure 5.** The dependence of the density radioactive contaminants in the stationary case (zero approximation) on the $z$ coordinate at different distances from the borehole: $1 - r = 0, 2 - 100, 3 - 200$. Other parameters: Ad = 0.01, Pd = $10^2$, $\delta = 10^{-4}, D_1^2 = 1$

Figure 5 shows a picture of the field distribution of the radioactive contaminant in the stationary case along the vertical coordinate (zero approximation). "Slices" are given for distances 0, 100$h$ and 200$h$ from the axis of borehall. We see that for the middle half-life nuclides ($T_{1/2} \sim 30$ years) in the covering and underlying layers pollutant density decreases rapidly, and even at distances of 0,5$h$ are negligible.

In general, the increase in the parameter Pd (rate of the injection of the solutions) leads to the "elongation" of the graphic along the radial direction, reducing the At (which corresponds to an increase in the average lifetime of the nuclide) - to "enhance" the graph along the axes $r$ and $z$. The field of pollutant remains limited.

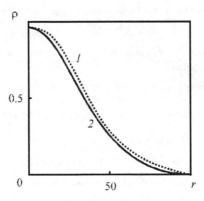

**Figure 6.** The dependence of the density of radioactive contaminant on the distance to the borehole axis Graphs are constructed (for the dimensionless time t = 100): the grid method - *1* and the method of asymptotic expansion - *2*. Other parameters: At = 0.1, Pd = 102, $\delta = 10^{-3}, D_1^2 = 1$

Fig. 6 shows the results obtained using a modified method of asymptotic expansions and the numerical solution to the problem of mass transfer by the grid method. Numerically was solved the problem (67) - (74), neglecting the radial diffusion.

Comparisons of the curves shown in Fig. 6 allow to conclude that the results obtained by numerical and asymptotic methods are in a good agreement.

So, based on the asymptotic method, approximated analytical solution to the problem of subterranean waste disposal is obtained, and accounting the boundary layer correction allows to provide the calculation of the areas of radioactive contaminants in the subterranean horizons with high accuracy at all distances from the injection borehall, and thus to clarify the forecast of the areas of radioactive contamination to ensure the environmental safety.

In conclusion, note that the above modification of the asymptotic method is quite general and provides the construction of "exact on the average" analytical solutions as to the nonstationary problem of the underground waste disposal as well as to the other problems of underground thermo- and hydrodynamics. The zero approximation of the asymptotic solution is of the particular importance, because it describes the average value of the variables, which is important for many practical problems.

## List of designations

| | | |
|---|---|---|
| A, B, C, E, F,M | – | auxiliary functions; |
| $a$, Ad, At | – | dimensional and dimensionless constants for radioactive decay for the diffusion and the temperature problem, respectively, 1/c; |
| $\lambda_{z1}$, $\lambda_z$, $\lambda_{z2}$, $\lambda_{r1}$, $\lambda_r$, $\lambda_{r2}$ | | coefficients of thermal conduction for the covering, porous, and the underlying layers in the vertical and radial directions, respectively, W/(m·K); |
| $\rho_n$, $\rho_{n1}$, $\rho_{n2}$ | – | density of the porous, covering and underlying layers, kg/m³; |
| $\rho_d$, $\rho_{1d}$, $\rho_{2d}$ | – | dimensional concentration of the impurity in a porous, covering and underlying layers, kg/m³; |
| $\rho_f$, $\rho_s$ | | dimensionless densities of the impurity in the carrier, the skeleton; |
| $D_r$,$D_{1r}$, $D_2$, $D_z$,$D_{1z}$, $D_{2z}$ | – | coefficients of diffusion of the layers in the radial and vertical directions, m²/s; |
| $\delta_{ij}$ | – | the Kronecker delta, |
| $\mu_s$, $\mu_w$ | – | chemical potentials of the skeleton and water, respectively |
| $g(\mu_s, \mu_w)$ | – | function of the mass transfer between the skeleton and the fluid; |
| H, h | – | power and half-thickness of the porous layer, m; |
| $K_r$ | – | Henry's coefficient; |
| $\bar{L}$ | – | differential operator; |

| | | |
|---|---|---|
| Pd | – | analog of the parameter Peclet; |
| $R_p$ | – | radiuses of the radionuclide contamination; |
| p, s | – | parameters of the Laplace-Carson,; |
| $r_0$ | – | radius of the bonehole, m; |
| $r_d, z_d, r, z$ | – | dimensional and dimensionless cylindrical coordinates, m; |
| $\tau, t$ | – | dimensional and dimensionless time, s; |
| $m, m_0, m_1$ | – | effective, initial and maximum porosity; |
| v | – | filtration rate, m/s; |
| | – | rate of convective transport of contaminant in the porous layer, m/s; |
| w | – | the true velocity of the fluid, m/s; |
| $q_d, q$ | – | dimensional and dimensionless source function (mass, kg/(s·m³);) |
| $k, k_1, k_2$ | – | stoichiometric coefficients; |
| $\theta, \theta_1, \theta_2$ | – | the remainder term of the asymptotic expansion in a porous, covering and underlying layers. |

## Author details

Mikhaylov Pavel Nikonovich[1], Filippov Alexander Ivanovich[1] and
Mikhaylov Aleksey Pavlovich[2*]

*Address all correspondence to: a.p.mikhaylov@gmail.com

1 Institute of Applied Researches, Sterlitamak, Bashkortostan, Russia

2 Department of General and Applied Physics, Moscow Institute of Physics and Technology, Dolgoprudny, Russia

## References

[1] Barenblatt G. I., Entov V. M. Ryzhik V. M. The movement of liquids and gases in natural reservoirs // - Moscow: Nedra. 1984. 211 p.

[2] Bear J. Dynamics of fluids in porous media. New York: American Elsevier publ. 1967. – 764 p.

[3] Bear J. Hydraulics of groundwater. New York etc.: McGraw-Hill Inc. 1979. XIII. – 567 p.

[4]  Bear J., Bachmat Y. Introduction to modeling of transport phenomena in porous media / Dordrecht et al.: Kluwer. 1990. – 533 p.

[5]  Bachmat Y and Bear J. Mathematical formulation of transport phenomena in porous media. Proc. Int. Symp. of IAHR on the Fundamentals of Transport Phenomena in Porous Media, Guelph, Canada. 1972. 174–197.

[6]  Ilyushin A. A. Continuum Mechanics. - Moscow: Moscow State University. 1979. 288 p.

[7]  Case V. M. Convective heat and mass transfer. - Moscow: Energiya. 1972. 448 p.

[8]  Landau L. D., editor. Landau L. D. and Lifshitz E. M. Continuum mechanics - Moscow: Gostekhizdat. 1954. 795 p.

[9]  Nigmatulin R. I. Fundamentals of the mechanics of heterogeneous environments. - Moscow: Nauka. 1978. 336 p.

[10] Nicholaevskij V. N. Mechanics of porous and fractured environments. - Moscow: Nedra. 1984. – 232 p.

[11] Nicholaevskij V. N, Basniev K. S., Gorbunov A. T., Zotov G. A. The mechanics of saturated porous layers - Moscow: Nedra. 1970. 336 p.

[12] Sedov L. I., Continuum Mechanics. - Moscow: Nauka. 1994. Vol. 1 and 2.

[13] Prakash A. Radial dispersion through adsorbing porous media. Proc. Am. Soc. civ. Engrs, 102 (HY3). 1976. 379 – 396.

[14] Barmin A. A., Garagash D. I. On the filtration of the solution in a porous layer taking into account the adsorption of impurities on the skeleton // Mechanics of liquids and gases. 1994. - № 4. 97-110.

[15] Bondarev E. A. Convective diffusion in porous layers, with taking into account the phenomenon of adsorption / Bondarev E. A., Nikolaev V. N. // PMTF, 1962. Number 5. 128 - 134.

[16] Zhemzhurov M. L., Serebrynij G. Z. Analytical model of radionuclide migration in porous layers // Journal of Engineering Physics. 2003. - T. 76. - № 6. 146 - 150.

[17] Zhemzhurov M. L., Serebrynij G. Z. Two-dimensional convective diffusion of radioactive impurities with taking into account the sorption in porous layers // Journal of Engineering Physics. 2008. - T. 81. - № 3. 417 - 420.

[18] Venetsianov E. V., Rubenstein R. N. Sorption dynamics of fluids. - Moscow: Nauka. 1983. 237 p.

[19] Philip J. R. Flow through porous media. Ann. Rev. Fluid Mechan. 1970. – 2. 177 – 204.

[20] Mikhaylov P. N., Filippov A. I. Asymptotic solution of the temperature field in a well with accout for the radial-velocity distribution // Journal of Engineering Physics and Thermophysics. – 2005. – V. 78. – № 4. 87 – 90.

[21] Mikhaylov P. N., Filippov A. I., Filippov K. A., Bagautdinov R, Potapov A. Temperature Field in Oil-Gas Beds Exposed to the Action of an Acid // Journal of Engineering Physics and Thermophysics. – 2005. – V. 78. – № 2. 256 – 271.

[22] Collins R. Fluid flows through porous materials. - Moscow: Mir 1964. 350.

[23] Lebedev A. V. Assessment of the balance of groundwater. - Moscow: Nedra. 1989. 178 p.

[24] Lukner L., Shestakov V. M. Modeling the migration of groundwater - Moscow:Nedra.1986. 209 p.

[25] Swing R. E. Simulation techniques for multicomponent flows // Commun. Appl. Numer. Meth. 1988. – V.4. – № 3. 335 – 342.

[26] Belitsky, A. S., Orlova E. I. Protection of underground water from radioactive contamination - Moscow: Medicine. 1969. - 209 p.

[27] Rybalchenko A. I., Pimenov M. K., Kostin P. P. et al., Deep burial of the radioactive waste, Moscow IzdAT. 1994. – 256 p.

[28] Sox M. D., Noskov M. D., Istomin A. D., Kessler A., Zubkov A. A. and Zakharova E. V. Modeling the distribution of radionuclides in the reservoir during deep burial of acidic liquid waste // Radiochemistry. 2007. - T. 49. - № 2. - 182 - 187.

[29] Larin V. K., Zubkov A. A., Kesler A. G. Modeling of the distribution of radionuclides in the reservoir during the deep burial of the acidic liquid waste. // Atomic energy. 2002. - T. 92. - № 6. 451 - 455.

[30] Kosareva I. M., Savushkina M. K., Arkhipova M. M., Wolin M., Kabakchi S. A., Egorov N. N., Rakow N. A., Kudryavtsev E. G. Temperature field in the deep disposal of liquid radioactive waste // Atomic Energy, 1998. - T. 85. - № 6. - 441 - 448.

[31] Kosareva I. M., Savushkina M. K., Arkhipova M. M., Wolin M., Kabakchi S. A., Egorov N. N., Rakow N. A., Kudryavtsev E. G. Temperature field in the deep disposal of liquid radioactive waste: modeling multi-step removal // Atomic Energy. 2000. - T. 89. - № 6. - 435 - 440.

[32] Kosarev I. M., Savushkina M. K. The temperature field in the deep disposal of liquid radioactive waste // Atomic Energy, 1998. - T. 85. - № 6. 441 - 448.

[33] Kosarev I. M., Savushkina M. K. The temperature field in the deep disposal of liquid radioactive waste: modeling of multi-stage disposal // Atomic energy. 2000. - T. 89. - № 6. 435 - 440.

[34]  Rybalchenko A. I., Pimenov M. K., Kostin P. P. The deep disposal of liquid radioactive waste - Moscow IzdAT. 1994. 256 p.

[35]  Chekalyuk E. B. Fundamentals piezometry of oil and gas deposits.- Kiev: GITL UkrSSR. 1965.-286 p.

[36]  Mikhaylov P. N., Filippov A. I., Guenther D. A , Ivanov D. V. "Exact on the average" asymptotic solution of the problem of underground radioactive waste disposal / Journal of Kherson National Technical University. B. 2 (28). - Kherson: KNTU. 2007. 365 - 370.

[37]  Mikhaylov P. N., Filippov A. I., Mickhaylichenko I. N., editor. Sabitov C. B. Field of concentration in the injection of fluid solutions of radioactive impurities in the deep layers / / Modern problems of physics and mathematics. Proceedings of the Russian Scientific Conference (16 - 18 September 2004, Sterlitamak) - Ufa: Guillem. 2004. 89 - 97.

[38]  Gurov K. P. Phenomenological thermodynamics of irreversible processes. - Moscow: Nauka. 1978. 128 p.

[39]  Mikhaylov P. N., Filippov A. I., Ivanov D. I. "Exact on the average" asymptotic solution to the problem of mass transfer during the underground disposal of liquid radioactive waste / Differential Equations and Related Topics: Proceedings of the International Conference (24 - 28 June 2008, Sterlitamak). - T. III. - Ufa: Guillem. 2008. 238 - 258.

[40]  Mikhaylov P. N., Filippov A. I., Guenther D. A., Ivanov D. V. The asymptotic solution to the problem of the underground disposal of radioactive waste // Siberian Journal of Industrial Mathematics. 2008. - T. XI. - № 2 (34). 124 - 138.

[41]  Mikhaylov P. N., Filippov A. I. The concept of "Exact on the average" asymptotic solution to the problem of the underground disposal of nuclear waste / SamDif 2007: Conference "Differential Equations and Their Applications", Samara, 29 - February 2, 2007 Abstracts. - Samara Universe groups. - 2007. 149 - 150.

[42]  Mikhaylov P. N., Filippov A. I., Guenther D. A., Ivanov D. V. On the construction of asymptotic solutions to the problems of conjugation/ / Journal of Computational Mathematics and Mathematical Physics. 2008. - T. 48. - № 11. 2046 - 2057.

[43]  Mikhaylov P. N., Filippov A. I., Mickhaylichenko I. N. Determination of the contamination zones in the underground disposal of solutions of radioactive dissolved substances // Bulletin of Kherson National Technical University. B. 2 (25). - Kherson: KSNTU. 2006. 508 - 511.

[44]  Mikhaylov P. N., Filippov A. I., Fattakhov R. G., Ivanov D. V.. Garifullin R. N., Guenther D. A. Boundary layer solution to the problem of mass transport of radioactive contaminants // Eurasian integration processes in science, education and produc-

tion: Proceedings of the Russian Scientific Conference (Kumertau, 19-20 October 2006) - Ufa: Guillem. 2006. 177 - 186.

[45] Mikhaylov P. N., Filippov A. I., Ivanov D. A., Guenther D. A. Temperature and concentration fields in the deep disposal of radioactive multicomponent solutions // Modern science: research, ideas, results, and technology. № 1, 2009. Series: Actual problems of thermophysics and physical hydrodynamics. Special issue based on the 7th Conference 21 - 25 September 2009, Alushta. 75 - 76.

[46] Mikhaylov P. N., Filippov A. I., Guenther D. A. The calculation of interrelated fields of concentration and temperature of fluids in the injection of radioactive solutions into the porous layer // Review of Industrial and Applied Mathematics. 2007. - T. 14. - V. 4. 754 - 755.

[47] Filippov A. I., Mikhaylov P. N., Mickhaylichenko I. N. Field of concentration in the injection of aqueous solutions of radioactive impurities in the deep layers // Review of Industrial and Applied Mathematics. 2004. - T. 11. - V. 3. 595 - 596.

[48] Filippov A. I., Mikhailov P. N., Guenther D. A., Ivanov D. V. Fields of concentration of radioactive materials // Journal of Engineering Physics. 2008. - T. 81. - № 5 912 - 923.

[49] Filippov A. I, Mickhaylichenko I. N., Guenther D. A. Construction of the "exact on the average" of the asymptotic solution of the stationary problem of filtration of radioactive solutions // Proceedings of the Sterlitamak branch of Bashkortostan Academy of Sciences. A series of "Physical-mathematical and engineering sciences." B. 4. - Ufa: Guillem. 2006. 64 - 74.

[50] Filippov A. I., Mikhaylov P. N., Mickhaylichenko I. N., Krupinov A. G. The calculation of concentration fields in the underground disposal of solutions of radioactive dissolved substances // Ecological systems and devices. 2006. - № 5. 27 - 33.

[51] Ditkin V. A. and Prudnikov A. P. Reference guide to the operational calculation. - M.: High School, 1965. - 465 p.

# Permissions

The contributors of this book come from diverse backgrounds, making this book a truly international effort. This book will bring forth new frontiers with its revolutionizing research information and detailed analysis of the nascent developments around the world.

We would like to thank Hironori Nakajima, for lending his expertise to make the book truly unique. He has played a crucial role in the development of this book. Without his invaluable contribution this book wouldn't have been possible. He has made vital efforts to compile up to date information on the varied aspects of this subject to make this book a valuable addition to the collection of many professionals and students.

This book was conceptualized with the vision of imparting up-to-date information and advanced data in this field. To ensure the same, a matchless editorial board was set up. Every individual on the board went through rigorous rounds of assessment to prove their worth. After which they invested a large part of their time researching and compiling the most relevant data for our readers. Conferences and sessions were held from time to time between the editorial board and the contributing authors to present the data in the most comprehensible form. The editorial team has worked tirelessly to provide valuable and valid information to help people across the globe.

Every chapter published in this book has been scrutinized by our experts. Their significance has been extensively debated. The topics covered herein carry significant findings which will fuel the growth of the discipline. They may even be implemented as practical applications or may be referred to as a beginning point for another development. Chapters in this book were first published by InTech; hereby published with permission under the Creative Commons Attribution License or equivalent.

The editorial board has been involved in producing this book since its inception. They have spent rigorous hours researching and exploring the diverse topics which have resulted in the successful publishing of this book. They have passed on their knowledge of decades through this book. To expedite this challenging task, the publisher supported the team at every step. A small team of assistant editors was also appointed to further simplify the editing procedure and attain best results for the readers.

Our editorial team has been hand-picked from every corner of the world. Their multi-ethnicity adds dynamic inputs to the discussions which result in innovative outcomes. These outcomes are then further discussed with the researchers and contributors who give their valuable feedback and opinion regarding the same. The feedback is then collaborated with the researches and they are edited in a comprehensive manner to aid the understanding of the subject.

Apart from the editorial board, the designing team has also invested a significant amount of their time in understanding the subject and creating the most relevant covers. They scrutinized every image to scout for the most suitable representation of the subject and create an appropriate cover for the book.

The publishing team has been involved in this book since its early stages. They were actively engaged in every process, be it collecting the data, connecting with the contributors or procuring relevant information. The team has been an ardent support to the editorial, designing and production team. Their endless efforts to recruit the best for this project, has resulted in the accomplishment of this book. They are a veteran in the field of academics and their pool of knowledge is as vast as their experience in printing. Their expertise and guidance has proved useful at every step. Their uncompromising quality standards have made this book an exceptional effort. Their encouragement from time to time has been an inspiration for everyone.

The publisher and the editorial board hope that this book will prove to be a valuable piece of knowledge for researchers, students, practitioners and scholars across the globe.

# List of Contributors

**Marek Solecki**
Department of Process Equipment, Faculty of Process and Environmental Engineering, Lodz, University of Technology, Poland

**Stanford Shateyi**
University of Venda, South Africa

**Sandile S. Motsa**
University of KwaZulu-Natal, South Africa

**Leôncio Diógenes T. Câmara, Flávio de Matos Silva, Guilherme Pereira de Oliveira and -Antônio J. Silva Neto**
Instituto Politécnico, Universidade do Estado do Rio de Janeiro, Brazil

**Jader Lugon Junior**
Instituto Federal de Educação, Ciência e Tecnologia Fluminense, Brazil

**Lídice Camps Echevarria and Orestes Llanes Santiago**
Facultad de Ingeniería Eléctrica - Instituto Superior Politécnico José Antonio Echeverría, Cuba

**A.M. Tsirlin and I.N. Grigorevsky**
The Program Systems Institute of RAS, Pereslavl – Zalessky, Russia

**Ahmed A. Khidir and Precious Sibanda**
School of Mathematics, Statistics and Computer Science, University of KwaZulu-Natal, South Africa

**M. I. Shilyaev and E. M. Khromova**
Department of heating and ventilation, Tomsk State University of Architecture and Building, Tomsk, Russia

**Sandile S. Motsa**
University of KwaZulu-Natal, South Africa

**Stanford Shateyi**
University of Venda, South Africa

**A. Abdelrasoul, H. Doan and A. Lohi**
Department of Chemical Engineering, Ryerson University, Victoria Street, Toronto, Ontario, Canada

**Mikhaylov Pavel Nikonovich and Filippov Alexander Ivanovich**
Institute of Applied Researches, Sterlitamak, Bashkortostan, Russia

**Mikhaylov Aleksey Pavlovich**
Department of General and Applied Physics, Moscow Institute of Physics and Technology, Dolgoprudny, Russia